Shipwrecks of Madagascar

Pierre van den Boogaerde

Strategic Book Publishing
New York, New York

Copyright 2009
All rights reserved — Pierre van den Boogaerde

No part of this book may be reproduced or transmitted in any form or by any means, graphic, electronic, or mechanical, including photocopying, recording, taping, or by any information storage retrieval system, without the permission, in writing, from the publisher.

Eloquent Books
An imprint of AEG Publishing Group
845 Third Avenue, 6th Floor — #6016
New York, NY 10022
www.StrategicBookPublishing.com

ISBN: 978-1-60693-494-4
SKU: 1-60693-494-5

Printed in the United States of America

To the people of Madagascar,

for having welcomed me during my three years of stay

there, I leave you a bit of your maritime history.

Contents

Introduction...1

Chapter 1—The Portuguese East India Company 21
São Vincente (1506)..23
Nossa Senhora da Luz and São Simão (1517) 29
Conceição and São Sebastião (1527) 30
São Ildefonso (1527)......................................34
Flor da Rosa (1528) 36
São Paulo (1538)..40
Esperança (Galega) (1540).................................40
Nossa Senhora da Barca (1559) 41
Nossa Senhora da Estrela (1660) 42
Nossa Senhora do Monte do Carmo (1774) 43

Chapter 2—The Dutch United East India Company 55
Alkmaar (1604)..62
Westfriesland (1606)..................................... 63
Brak (1613)...66
Gouda (1625) 67
Koning David (1639) 69
Tulp (1655)...72
Zeelt (1672) 74
Ridderschap van Holland (1694) 75

v

Chapter 3—The British East India Company 81
Anne (1689). 83
Degrave (1703) 89
Aurora (1770) 96
Winterton (1792). 97

Chapter 4—The French East India Company 113
Saint Alexis (1641) 117
Le Saint-Louis (1644) 119
L'Armand, Le Saint-Georges and La Duchesse (1656-57). 119
Taureau (1666) and Aigle-Blanc (1667). 120
Petit Saint-Jean (1670) and Saint-Luc (1671). 124
La Dunkerquoise (1674) 127
Soleil d'Orient (1681). 131
Le Vautour (1725). 135
Saint-Pierre and Salamec (1746). 137
Cerf and Phelypeaux (1757) 138
Gloire (1761). 140
Fortune (1775), Sirène (1776), Coureur and Indigent (1777). ... 141
Comte de Maurepas (1777). 144

Chapter 5—Pirates 151
Cygnet (1689) 158
John and Rebecca (1697) 164
Adventure Galley (1698). 165
Rouparelle (November) (1698). 173
New Soldado (1699) 173
Mocha (1699) 174
Bedford (1697) and Dolphin (1699) 176
Alexander (1700). 178
Charles, Buffalo, Rising Eagle and Dorothy (1708) and
    Neptune (1709) 180
Flying Dragon (1721) 182
Victorieux (1723). 183

Chapter 6—British Royal Navy. 197
Sibyl/Garland (1798). 198

Blenheim and Java (1807) and Harrier (1809) 198
Staunch (1811) 201

Chapter 7—French Navy 205
HMS Serapis (1781) 205
Chevrette (1830) 207
Colibri (1843) 209
Le Berceau (1846) 212
Lapérouse (1898) 213

Chapter 8—Sailing Vessels and Steamers of the 19th Century 217
Elisa (1827). 217
Margaret Oakley (1837) 218
Indienne (1852). 222
Macassar (1880) 225
Oise, Sarah-Hobart, Argo, Clémence and Armide (1885) 227
Surprise (1885) 228
Dayot, Belette, and Glide (1888) 233
Solitaire (1888). 234
SS Asiatic (Ambriz) (1903) 235
Peshawur (Ashruf) (1905) 236

Chapter 9—Russian Fleet 1904-1905 241
Russian transport ship (1905). 241

Chapter 10—Compagnie des Messageries Maritimes. . . 249
Le Tage (1890) 250
Douro (1910) 251
Salazie (1912). 251
Imerina (1932) 258
Bagdad (1935) 258
Amiral Pierre (1942). 259

Chapter 11—Compagnie Havraise Péninsulaire 261
Ville de Riposto (1899) 261
Ville d'Alger II (1920) 262
Catinat (1927) 263

Ville de Majunga (1928).................................. 265
Ville de Djibouti (1928) 266
Ville de Paris II (1935).................................. 267
Ville de Manakara III (1994) 268

Chapter 12—The Battle of Diego Suarez 271
Bougainville, HMS Auricula, Béveziers, Héros, and
    Monge (1942).. 271
M-16b and M-20b (Japanese Midget Submarines)............ 282

Chapter 13—Other Wrecks off Madagascar 287
Hirondelle (1743) 287
Glorieux (1755).. 288
Gange (1762)... 288
L'Heureux (1769) 289
Amphitrite (1799) 289
Chance (1799).. 289
La Désirée (1810) 289
Saint-Vincent de Paul, D'Après, and Meunier (1840) 290
L'Augustine (1855)....................................... 290
Anne-Marie (1882) 290
La Bourdonnais, Margareth and Irene (1893).............. 290
Conway Castle (1893)..................................... 291
Joseph A. Ropes (1894) 291
Lotsen (1894) 291
Myrtle M. (1897)... 291
Draguetta (1898)... 291
Alouette (1898).. 292
Espérance (1899)... 292
Falcon Hurst (1900) 292
Nina and Actio (1900).................................... 292
Roger (1901)... 293
Hermann (1901) 293
Princesse (1901) 293
Altaï (1902)... 294
Nansen (1902).. 294

Romford (1902) 294
Normand Macleod (and others) (1902) 294
Gertruda Gerarda (1903)................................296
Birmah (1903)..296
Marie-Thérèse (1904)..................................297
Bengalia (1905) 297
Cavalaire (Ivolina) (1926) 297
Pyrite (1926)...297
Beriziky, Sainte-Anne, Amanda, Alsace and Talisman (1927) ...297
Gudrun (1927) 298
Sitara (1929)...298
Albert-Morillon (1930).................................298
Mafia (1930)...298
Madina (1930) 299
Volontaire (1942).....................................299
Fort Franklin (1943)..................................299
U-Boot 197 (Submarine) (1943).........................299

Bibliography **303**
Archives 303
Books and Articles 304

Index..**325**

About the Author **335**

Acknowledgments

I wish to express my profound gratitude to all those without whom this work would never have seen the light. Their competence, friendliness, warm welcome and assistance greatly facilitated my research work. Their precious advice and unvarying support encouraged me to finish this book. I wish, in particular, to deeply thank the following people:

- Fred Lucas and Alexis Rosenfeld for having inspired and supported me since the beginning, and for our joint underwater discoveries and explorations in Madagascar.
- Madalena P. Vanoncellos for her invaluable research assistance of the Portuguese archives.
- Patrick Lizé for his support and precious research assistance.
- Robert Sténuit and Henri-Germain Delauze for their advice and support.
- Chantal Radimilahy, Director of the Antananarivo Art and Archeology Museum, for her valuable advice and boundless enthusiasm.
- The whole staff of the Bibliothèque Grandidier in Antananarivo, in particular Mrs. Olga Ranivoriaka, as well as the staff of the library of the Académie Malgache and of the National Archives in Antananarivo.
- Tim Healy for his infaillible support and his enthusiasm and precious assistance in reviewing the draft.
- Jean Philippe for all the information and pictures of the old Tamatave.

- Michel Pain for his support and precious assistance in reviewing and redrafting.
- Nirina Raharitsimba Andriamaherimanana for his precious assistance in preparing the maps.
- Vern Westgate for the superb editing job.
- The whole team at Strategic Book Publishing for their trust, encouragement and highly professional and most friendly advice.
- Dominique Buffin, Filipe Castro, Julien Durup, Eric Gilli, Paul Hawkins, Jean-Yves Le Lan, Charles Limonier, Philippe Murcia, Philippe Ramona, Jean-Aimé Rakotoarisoa, Colonel Franck Reignier, Chris Rule, Issop Sulliman, Serge Thebaut, for their assistance and support.

Finally, I wish to deeply thank my wife Catherine and my three children, Stephanie, Charles and Harold who suffered in silence for close to three years while I was engulfed in drafting this book.

§ § § § § §

To err is human. Notwithstanding my best efforts to compile all available sources, a number of mistakes or inaccuracies must remain. If you find some, make sure to let me know.

Introduction[1]

The island of Madagascar lies about 250 miles off the southeast coast of the African continent. The first sailing must have coincided with the arrival of humans on the island, which archaeologists place between 200 and 500 BC. Africans crossed the Mozambique Channel and settled along some of the coastline. However, a large number of the inhabitants of Madagascar are descended from Indonesians. They arrived on the island on outrigger canoes,[2] a journey of 3,700 miles, by following the eastern trade winds and the equatorial east-west current. Greek voyagers also made several journeys to the island.[3]

In the first millennium AD, Madagascar was visited regularly as part of a synchronized system of regular shipping between India, the Persian Gulf, the Red Sea and the east African coast.[4] The Indian Ocean has a predictable system of wind and sea conditions. Between November and March the winds blow from the northeast. From May to September the winds reverse and blow from the southwest. The currents conveniently follow suit. As a result, journeys between India and the Arabian Gulf down the east coast of Africa are reversible within six-month periods. Visiting vessels conveyed a wide spectrum of influences to Madagascar. They were not necessarily culturally uniform but were cosmopolitan, multi-ethnic homes to crews recruited for lengthy voyages.

Despite having little surety in the way of charts, the Arabs, recognized as skilled sailors, dominated trade in the Indian Ocean from about the 7th century until the arrival of the Portuguese in the early 16th century. Arab traders traveled down the east coast of Africa, crossed over to Madagascar, and established a series of trading posts which exploited inland resources largely by collecting goods which were assembled at coastal depots. These trading posts were concentrated in Madagascar's northwestern corner and along the west coast. Notably in the Bay of Boeny, Mahajanga and the islet of Nosy Manja at the entrance of the Bay of Mahajamba. Dhows from Malindi and Mombassa came to trade gold, silver and fine cloth from Africa, the Arabic peninsula and India for rice and slaves.

These Arab mariners, known in Malagasy as 'Antalaotse,' though mostly coming to trade before disappearing again, nonetheless left a strong strain still evident on the island. We note the large number of words in the Malagasy tongue borrowed from Arabic. These traders also spread the word of Islam. From the Mahajanga region, they went up the northwest coast, rounded the Amber cape, and plied along the coast following the eastern shore all the way down to Fort Dauphin.[5] Apparently, these Muslim traders or missionaries infiltrated during a relatively long period that started around the 10th century. Some Muslim immigrants settled in the southeast and rather superficially converted some of the local tribes to Islam. They then adopted the Arabic alphabet and geomancy.

The Chinese also had trade relations with countries of the Indian Ocean. In 1405, the grand eunuch Cheng Ho launched the first of seven large naval expeditions to the Indian Ocean. These were made up of seven fleets of more than 100 vessels each and spanned a twenty-year period.[6] They visited many countries around the Persian Gulf, the Maldives and along Africa's east coast. It is uncertain whether they landed at Madagascar. Generally though, Madagascar's inhabitants lived in relative isolation until French settlers started inland explorations in the second half of the 17th and 18th centuries.

European interests started with the great age of exploration by the Portuguese, Dutch, British and French fleets from the 16th century onwards. These explorers and merchants regarded the island as a strategically important site as it stood near the East-West trade routes and could be used as a way station on the journey to and from Asia. During that era, a voyage from Europe to Madagascar lasted around six months and longer if caught by the absence of winds around the equator or pushed by adverse winds and currents to the coast of Brazil. Ships in those days were very small by modern standards. Often more than a hundred crew and passengers were cramped in very close quarters and horrific hygienic conditions. Any outbreak of illness or fevers spread very quickly. In the absence of refrigeration, food and water reserves spoiled quickly. The lack of fresh produce resulted in most crew and passengers suffering from scurvy. The death toll on a voyage between Europe and India was frequently one third of the total crew and passengers, sometimes more.

Madagascar is a land of high plateaus, steep escarpments and arid plains. The eastern edge of the island has a mountainous spine which is frequently shrouded in mists that catch the water-laden trade winds after their long sweep across the Indian Ocean. The west coast of the island boasts many good harbors, innumerable coves with beaches that are ideal for careening ships, ample fresh water and provisions, including beef, chicken, eggs and citrus fruits such as limes and oranges, essential in the prevention of scurvy on long voyages.

In the ensuing 400 years, the European powers vied for influence over the island. They tried and often failed to establish settlements. The Portuguese established a small fort on an island in the Ranofotsy Cove[7] in 1528 but were massacred by the local population. They explored and signed treaties with chiefs in western Madagascar in 1613[8] and a Jesuit mission went up the river Manambovo to Sadia three years later. The Portuguese claimed western Madagascar in the Luso-Dutch treaty of 1641. An English settlement, set up in 1644 by a group of merchants led by William Courteen and Thomas Kynnaston at St. Augustine Bay near Toliary (Tulear), was terminated in 1646 because of fever, dysentery, hostile Malagasy tribes' people and the trying arid climate of southern Madagascar. Another English settlement in the northern island of Nosy Be came to an end in 1649. The French built a settlement at Taolagnaro, Fort Dauphin, in 1643. It lasted for about 30 years. It was abandoned in 1674 because of internal dissent, squabbles with the surrounding native population and neglect by the French East India Company.

While the Europeans' attempts to colonize proved unsuccessful, trading with the local chieftains and their tribes became very lucrative—for European and American traders alike.[9] The local chiefs were eager to trade slaves they had captured in battles with other tribes for gunpowder and firearms. The advantage of firepower helped native warlords capture even more slaves than before. Half of the number of captives were often traded with the arms merchants. Slave prices were very low because the market was outside the mainstream African slave trade. This made the long journey from Europe and the Americas worthwhile.

The Portuguese, who claimed Western Madagascar in 1641, used it mostly to supply slaves to the Dutch East India Company which had taken over Mauritius for its timber. A census taken at Barbados in the West Indies in 1680 found that half of their 32,473 slaves were from

Figure 1. Madagascar according to the Globe of Martin Behaim (1492), based on Marco Polo's fanciful descriptions and placing Zanzibar south of Madagascar (reproduced with the kind permission of Bibliothèque Grandidier, Antananarivo)

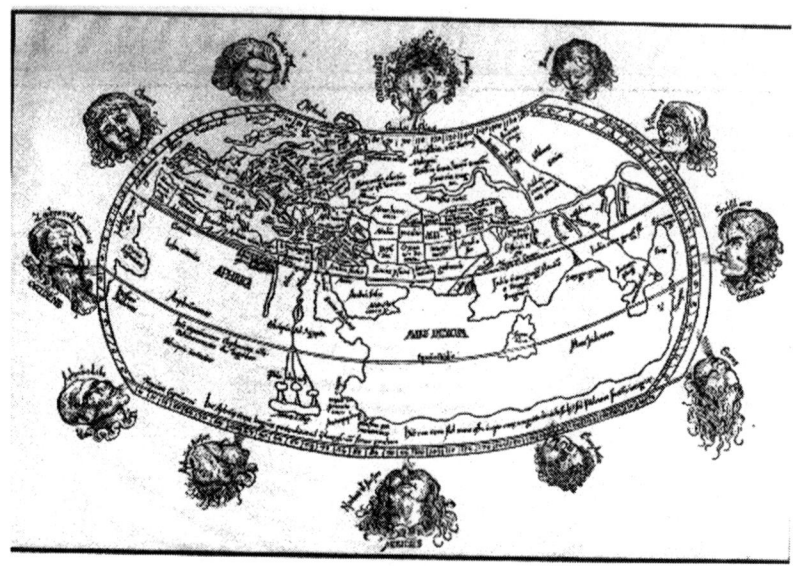

Figure 2. Map of Madagascar late 15th Century (ed. Ptolemy, 1483 and Reysch, 1503 in Grandidier, T. 1, p. viii bis)

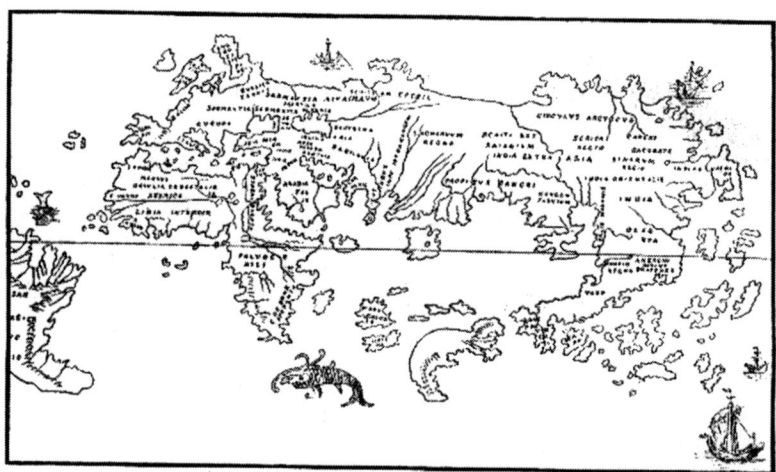

Figure 3. Map of Madagascar early 16th Century (Globe Lenos, 1510, in Grandidier, T. 1, p. viii bis)

Madagascar.[10] In the 18th century Madagascar supplied thousands of slaves to the French plantations on Mauritius and Bourbon (Reunion). In a reversal, the Merina kingdom imported slaves from the African mainland after consolidating their power around Antananarivo in the late 18th century.

Between 1680 and 1725 Madagascar became a pirate stronghold. Many pirates, having once visited, returned to settle down notably at Antongil Bay and Nosy Boraha, the Island of St. Marie, on the northeast side of Madagascar. They used these as safe ports from which they sailed and preyed on the rich trade of the Indian Ocean, the Red Sea and the Persian Gulf. There was even an attempt to establish a pirate-state in Diego-Suarez in the North of Madagascar called Libertalia under the leadership of Captain James Misson in the late 1600s.

By 1824, the Merina kingdom in the central highlands of Madagascar, a state of rice farmers who lived in relative isolation from the rest of the country for several centuries, had conquered nearly all of Madagascar. Notably, Ramboasalama had overthrown his uncle Ambohimanga about 1785 and had proclaimed himself King under the name Andrianampoinimerina. By 1792 he had eliminated two other local kings, united the Merina kingdom and moved his capital to Antananarivo. He built the royal palace, or *rova,* on a hilltop overlooking the city. By the time of his death in 1810, he had conquered the Bara and Betsileo highland tribes and was preparing to push the boundaries of his kingdom to the shores of the island.

His son Radama I (1792-1828) continued the expansion. He signed treaties with the United Kingdom outlawing the slave-trade and admitting Protestant missionaries into Madagascar. In 1824, having defeated the Betsimisaraka, Radama I declared, "Today, the whole island is mine! Madagascar has but one master." He was succeeded by his widow; Queen Ranavalona I (1828-61) nicknamed the Cruel. She murdered the dead king's heir and relatives, repudiated the treaties that Radama I had signed with Britain, persecuted Christian converts and asserted her power with the help of the aristocrats and sorcerers.

Her son and heir, the crown prince and future Radama II, grew up under the influence of French nationals in Antananarivo. In 1854, he wrote a letter to Napoléon III inviting France to invade Madagascar. On June 28, 1855 he signed the Lambert Charter. This document gave Joseph-François Lambert, an enterprising French businessman who arrived in Madagascar only three weeks before, the exclusive right to exploit all minerals, forests and unoccupied land in Madagascar in exchange for a 10-percent royalty to be paid to the Merina monarchy. In years to come, the French would use the Lambert Charter and the prince's letter to Napoléon III to justify an invasion in 1883 and another in 1895. These were called the Franco-Hova Wars and

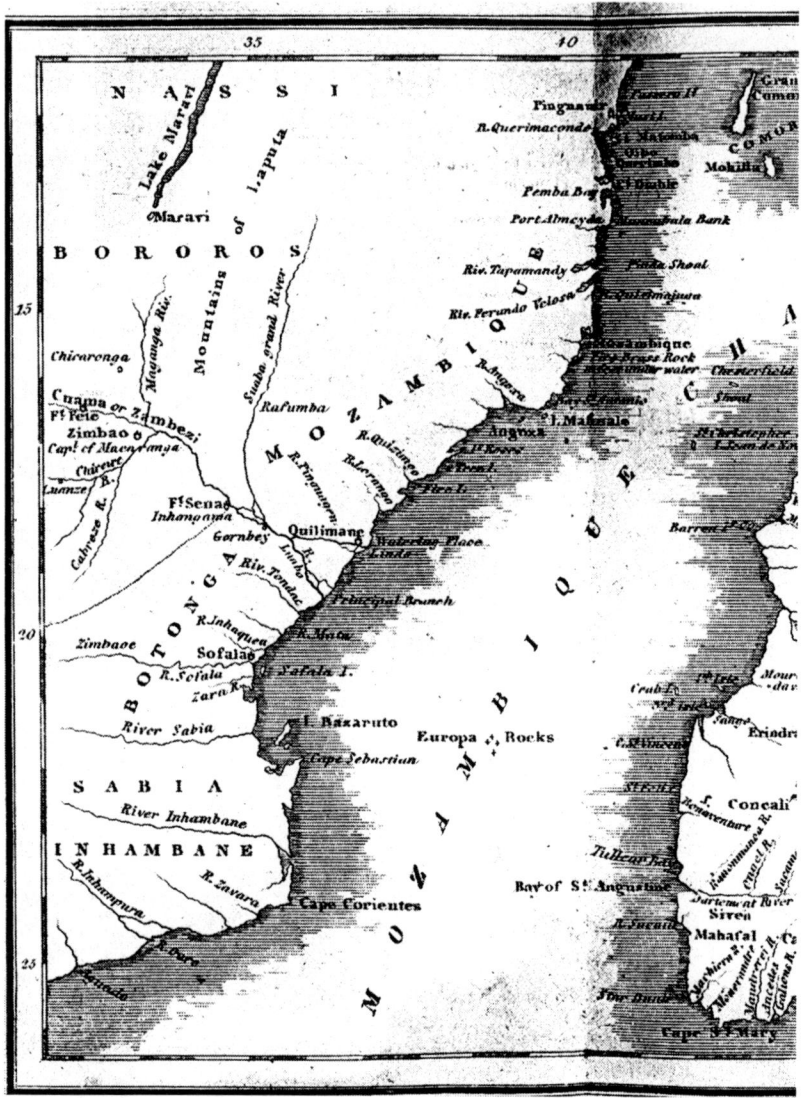

Figure 4. Map of Madagascar in the 18th Century

resulted in the annexation of Madagascar as a colony in 1896 at which time they put an end to the Merina monarchy.

After France fell to the Germans in the 2nd World War, the Vichy government administered Madagascar until 1942 when British troops occupied the strategic island to preclude its seizure by the Japanese. The Free French Forces received the island from the United Kingdom

in 1943. After suppressing the Madagascar revolt—a nationalist uprising—in 1947, France granted Madagascar full independence in 1960.

Given Madagascar's insular position, maritime links have been at the heart of its often turbulent history. Evidently, its treacherous waters and occasional cyclones resulted in a large number of ship-

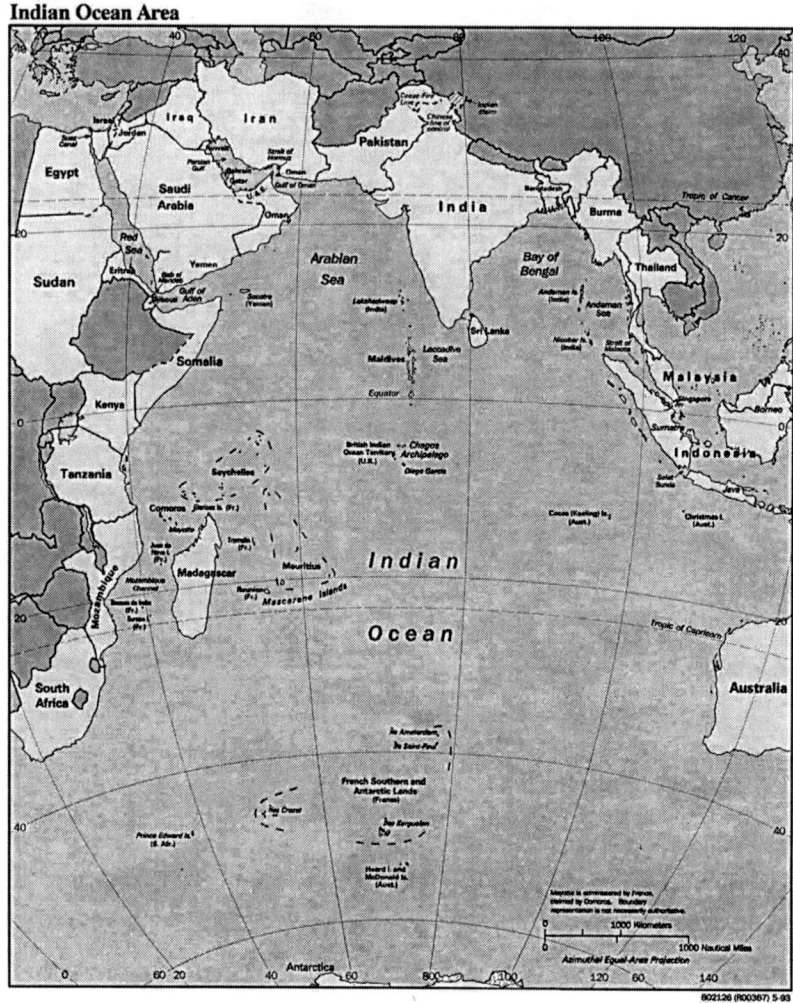

Figure 5. Contemporary map of the Indian Ocean (Courtesy of the University of Texas Libraries, The University of Texas at Austin)

wrecks over the centuries. These were caused by storms, navigation errors, decayed state of certain ships and naval battles. Unfortunately, documentation on wrecks of ships that sailed in ancient times from Indonesia, India, China, Java, the African mainland, Greece and the Arabic peninsula is virtually nonexistent. Remnants of an Indonesian wreck were found in the Bay of Antongil. An Arabico-Malagasy manuscript dating from the early 16th century, kept at the Bibliothèque Nationale in Paris, mentions that an Arabic dhow foundered on Madagascar's southeast coast, close to the mouth of the Matitana river, around the 7th century. There are fragmentary descriptions of a wreckage of a ship from Gujarat (northern India) in Ranofotsi Cove in the South of Madagascar in the first half of the 14th century; of the foundering in the 14th century of an Arabic dhow in the south of Madagascar on the return journey from a visit to southern Africa; and of an Indian ship from the Gulf of Cambaye that wrecked around 1480 close to Matitanana on Madagascar's southeast coast while sailing from India to Sofala (Mozambique). In other words, only a few specks compared to the numerous wreckages that must have occurred.

This book recounts the story of about a hundred notable shipwrecks off Madagascar and the fate and adventures of survivors. We cover ships of the Portuguese, Dutch, British and French East India Companies, of numerous pirates who visited or settled there, of the British and French Navies, of the sailing vessels and steamers of the 19th century, of the Russian squadron of 1904-05, of the steamers of the Messageries Maritimes and the Compagnie Havraise Péninsulaire, of the invasion by the allied forces in 1942 and of more recent times.

The map (Fig. 6, overleaf) shows the geographical location of these wreckages. Many of the early wreckages took place along Madagascar's southwest coast or in the south where these early navigators usually first caught sight of Madagascar. Either hugging the coastline or trying to anchor to get water and provisions, and in the absence of proper charts, the ships sailed onto Madagascar's numerous submerged reefs or were caught in ferocious cyclones. Obviously, many French ships foundered in the Fort Dauphin area given the French presence in that area. By the same token, most of pirate ship wreckages occurred on the east coast in the area of St. Marie Island, Foulpointe or Fenerive where the pirates had settled. With the development of Tamatave as Madagascar's main port since the late 19th century, most recent wreckages took place on the east coast.

Some of these numerous wreck have been explored but many still beg to be discovered.

Before relating the wreckages, I will present a short description of the customary interactions between European ships and the population of Madagascar at the time.[11] Until the 19th century, Madagascar was governed by a large number of local kings who constantly battled each other. Small fishing villages dotted the coast but the coastal royal towns or villages tended to be a little inland. When a ship anchored into a harbor, bay or cove they fired a cannon shot. In

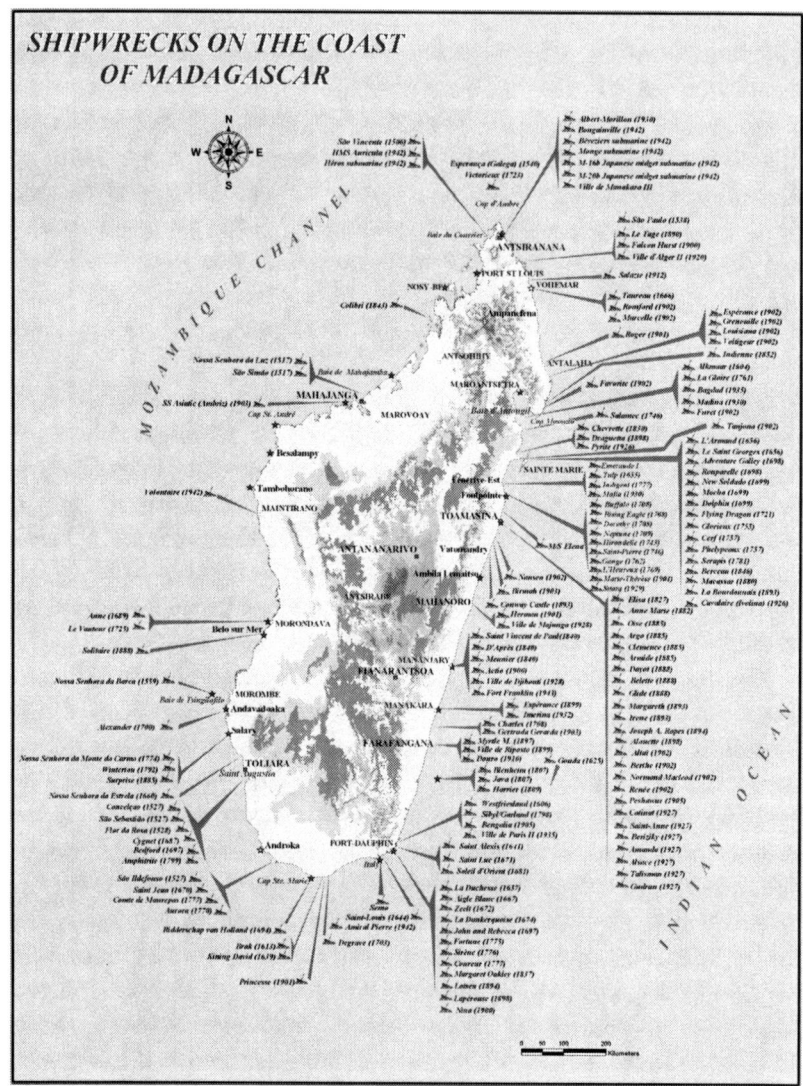

Figure 6. Geographical location of the wreckages

response the beach master lit a fire on the coast. This fire was repeated in succession inland so that within the hour the local king was informed that a ship had moored in the harbor. As soon as the ship anchored one or two natives came on board to inquire where she was coming from and which nationality she belonged to. Three to four members of the crew then went to the King's court with them carrying presents, usually weapons and cloth, requesting permission

Figure 7. Wrecks stranded on the beach at Fort Dauphin

Figure 8. Wreck in the harbor of Diego Suarez

to come ashore, fetch water, wood and other necessities, trade for provisions, slaves in the older days and, if needed, help in careening. Without permission, no one was allowed on shore and the local population was forbidden to come near the foreigners and trade. Sometimes the King would ask for assistance in fighting neighbors. When permission was granted, the King sent presents back, usually oxen, rice, fresh fruits and vegetables. Tents were put up, men took ladies for the duration of the stay and trade started in earnest. Though petty theft was common, the unconditional rule was that no harm could be done to the local population.

Stranded survivors of a wreck usually reported on the friendliness of the Malagasy, their compassion at times of distress and their interest in other cultures and ways of life. This said, surviving passengers had basically two choices; either live in the royal town or village and integrate or live along the coast waiting for a passing ship to take them along. Joining the community had the benefit of being fed, receiving a wife from the King, land and slaves. But return was rarely possible. One was linked through bondage, asked to contribute to the community's work (often teaching) and had to be ready to participate in wars against neighbors. Survivors who choose to stay close to the beach, while still under the protection of the King, had to fend for themselves or trade. However, beached survivors rarely had belongings in their possession. The security risk was much greater as the King was at a distance. If distance allowed, a third option was to walk to a larger port or bay where stranded survivors had a much higher possibility of encountering other Europeans and passing ships.

Myth and Reality

The 16th century discovery of the islands of the Indian Ocean and often exaggerated tales of returning seafarers and explorers caught the wildest imaginations early on. In 1694, Dutch writer Gerrit van Spaan published a fictional story[1] about Joris Pines and four women surviving the wreckage of a British ship *Indian Merchant* on an uninhabited island close to Madagascar. He wrote of their life on the island for more than 50 years and their 1,789 offspring! The story goes as follows. Four ships commissioned by British merchants sailed in April 1589 bound for India. After rounding the Cape of Good Hope and shortly before reaching Madagascar they were beset by a terrible storm. The four vessels got separated and the *Indian Merchant* was tossed around in the furious waters. One morning the ship was inexorably dragged toward a rock-strewn land. When nearing shore, the

Figure 9. Trade between European seafarers and Malagasy inhabitants (Source: Glazemaker, J.H. (trad.), "De Rampspoedige Scheepvaart der Franschen naar Oostindien onder 't beleit van Generaal Augustyn van Beaulieu, met drie Schepen, uit Normandyen," Amsterdam: Jan Rieuwertersz en Pieter Arentsz, 1669, fol. 24)

crew swung out the boat in which many passengers climbed while others jumped overboard to try to swim ashore. Only five people remained aboard: the accountant Joris Pines, the Captain's daughter, two maids and an African girl. They saw the boat capsize and most people drowned. The ship was violently thrust against the reef and broke up. The five clung to the bowsprit and landed on the beach. They walked up and down the coast looking for other survivors but found none. At nightfall, they fell asleep exhausted. The next morning they saw that most of the cargo and luggage had washed ashore providing them with many necessities. They built tents with pieces of sail first. Then they used the carpenter's tools that washed ashore to build a hut close to the mouth of a small stream using wood that was plentiful. They explored and learned that it was an uninhabited island but found plenty of fruits, nuts, birds, ducks, wild goats and fish to eat. They lived for about six months as brothers and sisters while getting organized. Eventually, though, Joris was overcome by carnal desires. He first seduced one of the maids but the others soon realized that something was going on. He then had gallant adventures with the second maid, the Captain's daughter and finally, the African girl. Soon children were born, their number grew rapidly and Joris educated them all in English and in the Christian faith. When they reached 16, 17 or 18 years of age, the half brothers and sisters married and multiplied further. Several villages were set up. When he got older, Joris married the grandchildren but made sure that children having the same grandmother never intermarried. Joris grew old, lost three of his wives and when he was 78 the Captain's daughter passed away. He named his oldest son as king of the island. When nearing 90 years of age and feeling weak he assembled all his offspring one last time: they numbered 1,789. He died shortly thereafter and was buried with great ceremony next to his four wives.

Another fictional story was published in 1841 by Mr. Jacomy-Régnier.[2] He reportedly published notes from his great-uncle Antoine, second-in-command aboard the French ship *Madras* that wrecked around 1700 close to Cape St. Marie in the South of Madagascar. The survivors decided to march to Fort Dauphin where they hoped to encounter Frenchmen and on the way met local inhabitants who spoke their Auvergne region dialect. The tale goes as follows: Around 1700, the *Madras*, bound for Pondichery, rounded Cape of Good Hope and was on its way to the inner passage through the Mozambique Channel. When sighting Cape St. Marie, the Captain had misgivings about a build up of clouds. He ordered all the topsails furled and closed all hatchways and scuttles. In an instant, frightening dark clouds surrounded the ship and a ferocious storm erupted. The violent winds broke the top gallant mast and pushed the ship broadside towards shore. To lighten the *Madras* from the water intake from the gigantic waves the crew cut down all the masts and threw the cargo overboard but to no avail. About a mile from shore her starboard side rammed and got stuck on a protruding rock springing a huge leak and causing a significant water intake. The pounding surf washed away the men who had neglected to fasten themselves and was cracking the timber. The ship was lost. The remaining crew members on board jumped and were washed ashore. Fourteen of a crew of seventy five survived.

Knowing that the local population could be hostile, the survivors armed themselves with slings built from shreds of their clothing, long sticks and locally built arches. During their long march across the mountains they fought several battles with local tribes in which nine were killed including the Captain of the *Madras*. When ready to cross the last mountain chain they were surrounded by a group of hunters carrying guns. Overpowered, they surrendered. The hunters asked them questions in Malagasy which they did not understand. However, overhearing the hunters speak among themselves, Antoine was surprised to recognize a distorted variety of Auvergne dialect, his mother tongue.

Antoine spoke to them in that dialect. He identified their little band as fellow citizens of their fathers. At once, their status changed from prisoners to brethrens. The five survivors were taken to a large village whose houses were modeled on those of Auvergne. In the ensuing days, natives of neighboring villages also came to greet them. They told them that this settlement was set up more than fifty years earlier. In the early reign of Louis XIII (1610-1643) flattering tales of returning seamen extolling all the riches to be found there lead to a commercial venture that was set up to create a settlement in Madagascar. Famine was ravaging Auvergne at the time so agents of the company scoured that region in search of potential settlers. They enticed about 30 families, a total of about 150 people, to follow them across the oceans.

Two vessels sailed from Bayonne and brought the settlers to the south of Madagascar close to Fort Dauphin. Initially, they lived near the coast but illnesses and frequent battles against the natives soon compelled the colony to move up the mountains where the air is healthier and where they could set up better defenses. They found fertile land, plenty or wildlife to hunt and rivers full of fish. They built a village on top of a hill they called Petit-Clermont.[3] At first they fought several battles with neighboring kings. They defended themselves so heroically that gradually they were respected and loved by their neighbors. The colony prospered. By 1700 they numbered around 900 people living between five villages named after cities in Auvergne, such as Petit-Saint-Flour and Petit-Aurillac.

After about six months in the village and nursing one of the survivors who suffered long fevers back to health, it was time to go to Fort Dauphin. Antoine tried to entice some of the villagers to accompany him to Europe but they adamantly refused. To protect them against attacks by local tribes, fifty villagers escorted the five survivors for 40 leagues until they were close to Fort Dauphin but refused to go further. A short while later, the survivors boarded a Dutch vessel returning to Europe. Antoine sent a report of what he had found to the government, however, they did not follow up on it.

The publication in 1719 of Daniel Defoe's hugely successful novel "The Life and Strange Surprising Adventures of Robinson Crusoe of York, Mariner," which is regarded as the first realistic novel of the world literature made the genre very popular. This made it hard to separate myth from reality at times. The adventures of Robert Drury, who lived for 15 years in Madagascar after surviving the wreckage of the *Degrave* in 1703 (see below), though true, were long dismissed as fiction. They were attributed to Daniel Defoe as it came out ten years after he published Robinson Crusoe. Likewise, some still attribute the earliest collection on piracy, Captain Charles Johnson's "A General History of the Pyrates" published in 1724, to Daniel Defoe.

The fictional tradition lives on. A more recent work is Louis Garneray's "Les Naufragés du Saint-Antoine," vaguely inspired by Drury's tale and first published in 1951.[4] It tells the story of a fictional French ship that wrecks in Madagascar after a tumultuous chase with an English pirate and the heroic tales of the survivors to stay alive amidst a hostile population and environment.

Sources

[1] van Spaan, G., "De Gelukzoeker over Zee of D'Afrikaansche Weg-Wijzer," Chapter VII, Rotterdam: Pieter vander Slaat, 1694, pp. 69-99.

[2] Jacomy-Régnier, "Colonie auvergnate à Madagascar," Revue d'Auvergne, T. 1, March 1840-1841, pp. 166-174; and "A la recherche d'une colonie perdue," Marine de France, No. 6, January 25, 1895; see also: Meyniard, C., "A la recherche d'une colonie perdue-Les Auvergnats à Madagascar," Bulletin de la Société de Géographie de l'Est, 1905, pp. 52-55; and Galli, H., "La Guerre à Madagascar-Histoire Anecdotique des Expéditions françaises de 1885 à 1895," Paris: Garnier Frères, s.d., pp. 32-35.

[3] From the capital of the Auvergne region, Clermont Ferrand.

[4] Garneray, L., "Les Naufragés du Saint-Antoine," Paris: GP, 1951 (second edition Paris: GP, 1952, and third edition, Paris: L'ancre de Marine, 2002).

REFERENCES

[1] The material in this introduction is borrowed from many sources, but notably from Mack, J., "The land viewed from the sea," Azania, XLII, 2007; Vérin, P., "The History of Civilisation in North Madagascar," Rotterdam: Balkema, 1986; Grandidier, A., Charles-Roux, Delhorbe, C., Froidevaux, H., and Grandidier, G., "Collection des ouvrages anciens concernant Madagascar," Paris, Comité de Madagascar, 1903; and the excellent summaries in Wikipedia, History of Madagascar, http://en.wikipedia.org/wiki/History of_Madagascar; and Chris Rule, "Piratical History of Madagascar," in Pirate Strongholds & Hideouts, http://www.piratesinfo.com/detail/detail.php?article_id=70

[2] Dahl, O.C., "Migration from Kalimantan to Madagascar," Norwegian University Press / The Institute for Comparative Research in Human Culture (1991).

[3] See Grandidier, A., « Histoire de la Géographie de Madagascar », Paris, 1883, pp. 1 to 11. Ptolemy was the first to mention the island, though his picture of the world contained errors that affected navigation and discovery. He adopted a false estimate of the circumference of the earth, making it about one-sixth too small. He enclosed the Indian ocean in a continent, which extended from Africa to China, and said that the whole southern hemisphere was too hot for navigation (Gilbert, William, *Renaissance and Reformation.* Lawrence, KS: Carrie, 1998, Chapter 10, "Exploration and the discovery beginnings of the expansion of Europe," formatted and installed as an e–book by Lynn H. Nelson at the University of Kansas, at http://vlib.iue.it/carrie/texts/carrie_books/gilbert/10.html, p. 3)

[4] For more details, see Mack, J., op. cit.

[5] For more details, see Ferrand, G., "Les Tribus Musulmanes du Sud-Est de Madagascar," Revue de Madagascar, June 1903, pp. 481-491.

[6] For more details, see Godfrey, T., "Dive Maldives, A Guide to the Maldives Archipelago," Richmond (Victoria, Australia): Atoll, 1996, pp. 9-10.

[7] Islet of Anosy, also called Islet of the Portuguese, or Tranovato in Malagasy, off the town of Italy, west of Fort Dauphin on the southeast coast of Madagascar (see story below).

[8] Explorations by Captain Paulo Rodrigues da Costa and Jesuit fathers Pedro Freire and Luis Mariano aboard the caravel *Nossa Senhora da Esperança* in 1610-11 and 1613-14.

[9] Chris Rule, op. cit.

[10] Africa and Slavery 1500-1800, in Sanderson Beck, "Middle East & Africa to 1875," s.l., 2004 (also on the web at http://san.beck.org/1-13-Africa1500-1800.html#11)

[11] This section was inspired by a large number of accounts, but notably by de Bucquoy, J., "Zestien Jaarige Reize naar de Indiën gedaan door Jacob de Bucquoy, vol Aanmerkelyke Ontmoetingen," Haarlem: Jan Bosch, 1758 (second ed.), pp. 45-46 and 89-98.

1

The Portuguese East India Company[12]

The great age of European exploration and discovery was inaugurated by the Portuguese. The European nations along the Atlantic coastlines had grown in strength and were seeking new trade routes and new land as a way to counteract the long-standing Italian, and particularly Venetian, monopoly of the eastern trade. Economic impulses, including spices, precious metals and the slave trade were dominant but religious and missionary aspects were also present. The Portuguese came first by pioneering improved maritime technology. The clumsy square-rigged ships used in the 1400s that could only sail windward, were difficult to maneuver and unsuited for long journeys or adverse winds were replaced by the two-master lateen caravel used by Arab merchants. The lateen sail was more or less triangular and capable of being adjusted to almost all winds. The Portuguese improved the caravel by combining the square-rig with the lateen sails and adding a mast, or sometimes two.[13] As a result the advantages of both types of ships were combined and the new ships made long-distance voyages feasible.

The first important figure in history of exploration is Prince Henry the Navigator (1394-1460), a member of the Portuguese royal family. He sent out expeditions from his palace on the coast of Sagres including one to explore the coast of Africa in 1420. These voyages began continuous ocean sailing. The slave trade and gold discoveries assured continued support of African exploration after Henry's death. By sailing ever further south along the west coast of Africa, progress was made towards reaching the Southern tip of Africa, which was much farther than was realized. First to round the southernmost point of Africa was Bartolomeu Dias, sent by King João II in 1487 with two ships. Blown off course by a storm in the neighborhood of Walfish Bay, Dias reached the coast and rounded the tip without realizing it. He sailed eastward as far as the Great Fish River, five hundred miles beyond the Cape before turning back. On the return voyage he realized that he had rounded the southern tip of Africa.

A land expedition to India via Egypt and Yemen was sent by King João the next year (1489) under the leadership of Pro da Covilh. Having established that it was possible to sail east around Africa, Vasco da Gama left Lisbon in July 1497 with four ships and achieved the first successful trip to India around the Cape of Good Hope in May 1498. He returned to Lisbon in September 1499. He was gone over two years and lost a third of his men but Vasco da Gama had opened a maritime route to the Asian markets of spices and exotic goods. The Portuguese crown sought to exploit this new route and keep this commerce under state control. As a result, it sent fleets to India for an estimated total of 1,154 voyages. They were sent nearly every year between 1497 and 1700.[14, 15] The Portuguese trading season in the Indies lasted from September to April as harbors on the west coast of India were closed for the southwest monsoon. These armadas usually left Lisbon after Easter to round the Cape of Good Hope and catch the tail-end of the southwest monsoon winds off the east coast of Africa, ending up in Goa in September or October. The return journey aimed at leaving Goa or Cochin with the northeast monsoon.

The Portuguese first became aware of Madagascar when the explorer Diogo Dias was swept wide of the Cape of Good Hope and separated from the other ships of the fleet during a violent storm in May 1500. After the storm dissipated, Diogo drifted along the east coast of an island which he named Isle of Saint Laurence (São Lourenço)[16] as he first sighted it on August 10, 1500, this saint's name day. At first he thought he was sailing along the coast of Mozambique but when he reached the northernmost point of Madagascar he realized he had discovered an island. After staying on the island for a short while he crossed back over to Africa.[17]

Of an estimated 219 Portuguese shipwrecks between 1497 and 1650,[18] 58 (26.5 percent) were recorded lost in the Mozambique Channel.[19] About a dozen are documented to have wrecked off Madagascar. Another 5 or 6 naús of the early fleets (between 1500 and 1510) are believed to have foundered off Madagascar, but the evidence is sketchy. These include, possibly, three wreckages in 1504: the *Nossa Senhora da Conçeição* (Captain Pedro de Ataíde); the *Rainha* (Captain Francisco de Albuquerque); and the *Faial* or *Faia* (Captain Nicolau Coelho); and two wreckages in 1507 (the naús, names unknown, of João Rodrigues Pereyra and Rui Mendes). Thus, about one third of the total lost in the Mozambique channel foundered off Madagascar, the rest off Africa's east coast. This is explained by the fact that the naús taking the Mozambique Channel route to India generally tried to stay close to the East African coast giving them access to the Portuguese settlements in Mozambique, Zanzibar or Mombassa. Most of those that wrecked in Madagascar had gone astray because of storms or navigation errors. Also, most documented wrecks off Madagascar date from the first half of the 16th century. This is the early period of the Portuguese India trade when marine technology and performance at sea was not yet well developed.

SÃO VINCENTE (1506)

The *São Vincente*, under the command of Captain Rui Perreira de Coutinho, took part in the first Portuguese attempt to explore Madagascar.[20] She was part of a double fleet that left Lisbon on March 6, 1506: on the one hand, about 10 ships used for trade and exploration commanded by Tristan da Cunha on the admiral ship *São Iago;* on the other hand, four navy and conquest ships led by Alfonso de Albuquerque on the naú *Cirne*. The fleets were dispersed by a massive storm close to the Cape of Good Hope. The fleets eventually reassembled at the Primeiras islands[21] and then sailed to Mozambique, save for two ships: the *Garça*, captained by Alvaro Telos, after having trailed the east coast of Madagascar went to Malindi (in Kenya) to wait for the fleet at Cape Guardafui (Somalia); the other, the *São Vincente*, anchored in the harbor of Matitanana[22] in August, 1506.[23]

As soon as Captain Rui Perreira anchored, dugouts came to his ship and local inhabitants boarded and showed him ginger, cloves, other spices, wax, cotton cloth and silver shackle bracelets[24] claiming that vast quantities of these products could be had inland. He was tempted to explore these riches but the shipmaster and the pilot refused. He then embarked two of the natives and sailed to Mozambique to inform Tristan da Cunha of the potential vast resources that

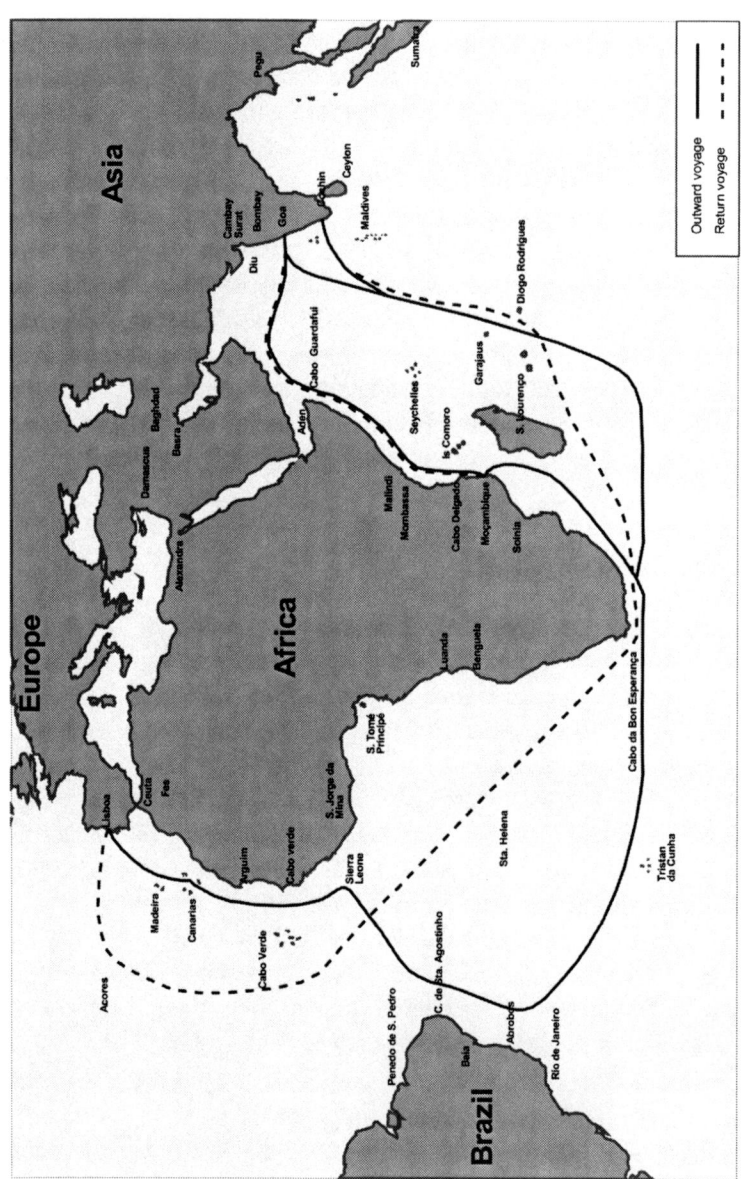

Figure 10. Portuguese India routes

Figure 11. Unidentified Portuguese ship wreckage in Madagascar (Source: De Bry, Icones tertice partis Indiæ orientalis, 1601, pl. II)

he had found including possibly precious ambergris. Tristan da Cunha was enthused by the news. He hoped that he had found a new India and that, after visiting the coast of Madagascar, he could send some of the ships back to Portugal with a rich load. He could not sail to India at that time due to the northeast monsoon. Instead of idle wintering in Mozambique Tristan da Cunha decided, against the advice of Alfonso de Albuquerque, to explore Madagascar with seven ships.[25] The pilots recommended circumnavigating Madagascar through the North.

The fleet left Mozambique in November 1506. After crossing the Mozambique Channel, it anchored in the Bay of Boeny in western Madagascar.[26] Tristan da Cunha sent a small forward party including a Malagasy interpreter he had found in Mozambique to shore. The first contact was amicable, but on the way to see the local king, the natives attacked the interpreter because he had brought Christians to the island. He owed his life to his Portuguese escort firing in the air. In light of this hostile conduct, Tristan da Cunha decide to leave the next day.

He sailed ten miles north to the entrance of the Bay of Bombetoke (Mahajanga) and made a friendly contact with the local population.

Ambergris (Grey amber)[1]

A main treasure that early explorers sought in Madagascar was the highly prized ambergris, these valuable nuggets that can be found floating in the sea or washed on the beaches of islands. Ambergris is a waxy substance produced in the intestines of sperm whales. Sperm whales eat deep sea squid and it is believed the undigested beaks of the squid cause irritation to the internal organs of the whale. In response, the whale produces a biliary secretion which coats these beaks and which is passed out at either end. When heated, it gives out a particularly agreeable fragrance, much like musk, that is used for medicinal purposes and as a fixative in perfumery. Given its cost, today natural grey amber is mostly replaced by synthetic substitutes.

Around the 10th century, the Arabs thought that grey amber was a plant growing at the bottom of the sea that was washed ashore during storms. Others believed that it was dung of large birds. The Chinese called ambergris 'dragon spit.' An unknown Portuguese found the true source of ambergris early in the 16th century. Observing inhabitants cutting dead whales washed ashore on the Laccadive islands,[2] he noticed that some of the whales had pieces of ambra like pinecones, big and small ones, sometimes numbering more than one hundred, joined in a sticky bunch. Tempted by immediate riches, explorers, starting with the Portuguese, were eager to find ambergris that could be sold to the royals and the very rich for a large profit.

Raw ambergris fetches approximately US$10 per gram (as of 2006), with much higher prices possible for particularly high-quality samples.

Sources
[1] Ambergris, Wikipedia at http://en.wikipedia.org/wiki/Ambergris and "The sea of the Maldives" at http://www.hellomaldives.com/maldives/sea/contents.htm.

[2] A group of coral reefs and islands in the Indian Ocean.

That night their chief guided him to the entrance of the well protected Bay of Mahajamba to a densely populated islet (Nosy Manja) where Arab traders had established trading posts and where the local king lived. Seeing Portuguese ships, the inhabitants of the islet panicked and tried to flee to the mainland. Some jumped into canoes and others attempted to swim. Many of the overloaded canoes capsized and the sea was soon covered with more than 200 corpses of men, women and children. To control the channel and prevent more inhabitants from fleeing, Tristan da Cunha placed two barges armed with petrero cannons at the entrance of the channel. He then anchored the ships in the harbor. Most of the men fled before the blockade. They found about 500 people on the island, mostly women and children. There were only about 20 men left, including the chief which they named the Cheikh.

After plundering the city and seizing a lot of gold, silver and rich fabrics the Portuguese spent the night feasting. At daybreak the next morning, canoes approached the island manned by about 600 men ready to die to get back their women and children. Tristan da Cunha handed them over, indicating that he had never intended to harm them but just came for provisions and for gathering information on what nature was producing in their land. This appeased the local population. They provided Tristan da Cunha with supplies for his onward voyage.

Tristan da Cunha followed the west coast on a northward course. He stopped off a densely populated town called Cada,[27] whose residents were mainly Arabs and marooned slaves who had escaped from cities on Africa's east coast. The reception was hostile so he burned the town and continued north. On Christmas day 1506 he reached the northernmost point which he called Cape Natal.[28] However, he was unable to round this cape because of adverse winds and currents. After losing time battling the elements, he granted Alfonso de Albuquerque's wish to return to Mozambique with four ships in order to proceed to India. Tristan da Cunha set sail on a southern course with the four other ships. He thought that by following Madagascar's western coast and rounding its southern point it would be easier to reach Matitanana with all its promises of riches.

Shortly after the fleet divided, during a stormy night with fierce winds and heavy seas, the *São Vincente* was hugging the coast ahead of the others. She was suddenly thrown against one of the many islets dotting the entrance to Courrier Bay. The ship was lost and Captain Rui Perreira died in the wreckage. Tristan da Cunha's *São Antonio,* which was very close, would have perished also were it not for the shipwrecked sailors of the *São Vincente* crying out to warn him of the danger, allowing the pilot to steer his ship away. The next morning, Tristan da Cunha was in the company of only one other ship. The *São Vincente* was lost and the *Nossa Senhora da Luz* had disappeared! He did not wish to continue sailing in such dangerous waters. He sailed to the Comoros islands where he met the rest of the fleet and returned to Mozambique.

Only the master, the pilot and 13 sailors survived the wreckage of the *São Vincente*. They managed to reach Mozambique in a barge. Tristan da Cunha sent them back in the *São Antonio* commanded by Captain João da Vega to find out whether it was possible to salvage the money chest. They did not succeed and João da Vega rejoined the fleet in Malindi (Kenya).

Before sailing to India, Tristan da Cunha sent the *Santa Maria* under the command of Antonin da Saldanha to Portugal with a letter to King Dom Manoel describing what he had done and discovered. He put the two Malagasy who had embarked with Captain Rui Perreira aboard and sent the silver shackle bracelet that Rui Perreira had received in Matitanana. The *Santa Maria* arrived in Lisbon in August 1507. King Manoel was excited by what he heard and decided that further exploration of Madagascar was called for.

Notwithstanding Tristan da Cunha's doubts, the *Nossa Senhora da Luz* had not disappeared. João Gomes de Abreu had successfully rounded the Amber Cape and, unaware of what was going on, reached the mouth of the river Matitanana and was waiting for the admiral ship to arrive.[29] About 20 dugouts soon surrounded the anchored ship, offering fish, fruits and sugar cane. The master, who spoke Arabic and several other languages, boarded one of the canoes and was taken away. Not seeing him come back, Abreu armed the yawl with petrero cannons and took 24 men to look for him. When he got close to the coast, he saw the dugouts coming back with the master covered in a local cotton garb, wearing silver bracelets and rings and having a large chain embedded with thirty silver crusades around his neck. All of which was given to him by the local king as a sign of friendship. Seeing this, Abreu decided to visit the king and had a meal with him. His meal barely over, a storm erupted, the race at the mouth of the river became impassable and he was stranded on land for four days. Not seeing their Captain come back, afraid of being thrown ashore by the storm and missing the pilot who had accompanied Abreu, the sailors set sail and reached Africa's coast close to the Angoza island (Southern Mozambique) where they met Ruy Soares. The two ships sailed together to Mozambique reaching it after Tristan da Cunha had already left.[30]

Abreu and eight of the 24 sailors who had accompanied him quickly died from tropical diseases. Thirteen sailors decided to try to reach Mozambique with the yawl preferring to die at sea than from illnesses on land. After reinforcing her and loading water and provisions, they left in early 1507, leaving the three remaining sailors in Matitanana. After following Madagascar's east coast and rounding Amber Cape, they stopped in one of the bays in the northwest to fetch water. There they were treacherously hit by assegais and stoned by local inhabitants. Though some were wounded, they managed to cross the Mozambique Channel and eventually arrived at the Angoza islands where they found Captain Lucas da Fonseca who was returning from India. Fonseca took them aboard his ship and brought them to Mozambique, from where they could reach Goa.

Two of the remaining three stranded sailors, a Portuguese named Andre and a Genoese named Bartholome, were picked up in August 1508 by Captains Diogo Lopez de Sequeira and Duarte de Lemos who had anchored in the Bay of São Sebastian[31] while exploring Madagascar. They found the third one, named Antonio, a little later in Fort Dauphin. Antonio had learned the Malagasy language and served as an interpreter to Diogo Lopez.

NOSSA SENHORA DA LUZ AND SÃO SIMÃO[32] (1517)

The armada departing from Lisbon in the spring of 1516, consisting of five naús and a navio (smaller ship), left in two batches. The naú *Nazaré* of the capitão mor (commander) João da Silveira left on March 22, 1516 with the *Nossa Senhora da Luz* captained by Francisco de Sousa "Mancias" and the *São Simão* commanded by Antonio de Lima. The two other naús, commanded by Alfonso Lopes da Costa and his brother Garcia da Costa, left on April 4 with a special message from the King to Alfonso de Albuquerque. Being skilled sailors, the two brothers arrived in Goa well ahead of the others to find Albuquerque already dead.

The first three naús had a rough trip. After being delayed in the Atlantic Ocean they finally made it to Mozambique. There, João da Silveira had orders waiting telling him to go meet the Governor in Quiloa.[33] The three naús sailed from Mozambique in early 1517 but encountered a furious storm. Unable to steer, they were pushed eastward toward Madagascar. The *Nazaré* miraculously survived the storm after cutting down her masts. The *Nossa Senhora da Luz* and the *São Simão* were thrown ashore in the Bay of São Lazaro (north of Mahajanga). The Captains and most of the crew survived the wreckages. When the storm subsided, Francisco de Sousa Mancias loaded the longboat with a few crew members and some of the cargo and went to the dreadfully damaged *Nazaré*. They proceeded to Quiloa. After wintering there and replacing her masts, the *Nazaré* proceeded with the coming of the monsoon to Goa, arriving in August 1517.

CONCEIÇÃO AND SÃO SEBASTIÃO[34] (1527)

On March 26, 1527 an armada of five ships left Portugal bound for India under the command of Captain Manoel de Lacerda on the admiral ship *Conceição*.[35] He was accompanied by the *São Sebastião*[36] captained by Aleixo de Abreu, the *Flor de la Ma*r, commanded by Balthazar da Silva, the *São Roque,* Captain Gaspar de Paiva, and a naú commanded by Christovão de Mendoza. After an uneventful voyage, the fleet reached southern Madagascar by mid-summer but drifted apart in the Mozambique Channel due to contrary winds. At night, in a choppy sea with strong currents and absent proper charts and accurate knowledge by the pilots the *Conceição* hit sandbanks about three leagues south of the Bay of Saint Augustine. The *São Sebastião* which closely trailed the poop lanthorn of his admiral ship by night did not have time to tack and also foundered on the sandbanks.[37]

The next morning the crews, soldiers, passengers, about 800 people overall managed to reach nearby land on rafts. For protection, they dug trenches in which they hid with some of the supplies, arms and other items that could be salvaged or that had washed ashore. They stayed close to the beach and did barter for food with the local population while hoping that a ship would eventually pass close by to pick them up. As this part of Madagascar is arid and poor they lived miserably and many died from diseases.

In early July 1528, about a year later, they saw a ship (the *Nossa Senhora de Ajuda* belonging to Nuno da Cunha's fleet and captained by António de Saldanha). As evening fell, they built big fires in the shape of a cross to indicate that there were Christian shipwreck survivors wishing to be rescued. António de Saldanha's crew saw the

Figure 12. The Portuguese Armada of 1516, showing the *Nossa Senhora da Luz* and the *São Simão* thrown on Madagascar's west coast (source: Bellec, F., Oliveira, R., and Michéa, H., "Naus, caravelas e galeões: na iconografia das descobertas A careira da Índia no século XVI: realto de uma vulgar viagem ao inferno A arquitectura naval e a expansão marítima portuguesa Princiais navios de século XVI," Lisboa: Quetzal, 1993, image No. 71, pp. 64-65.)

Figure 13. Details of the *Nossa Senhora da Luz* thrown on Madagascar's west coast

Figure 14. Details of the *São Simão* thrown on Madagascar's west coast

signs, furled the sails and waited. The next morning, they came closer to shore but kept some distance, being afraid to founder on the unknown shallows. They were a little uneasy thinking it might be a trap. They hoped that a dugout would come out to the naú to tell them who these people were. The survivors were waiting for one of the ship's boats. After waiting for eight days of approaching during daytime and retreating at night, they had to sail on due to bad weather, leaving the brokenhearted survivors ashore.

After despairing of being saved in St. Augustine, Manuel de Lacerda, Aleixo de Abreu and their crews decided to walk overland to Matitanana, more than 500 kilometers away on the other side of Madagascar. They hoped that food would be more plentiful than in the barren area where they were and where they thought they had a better chance of being rescued or finding a barge to sail to Mozambique. They left behind one sick sailor, who was picked up by Captain Nuno da Cunha in August 1528, just before the *Flor de Rosa* wrecked in the same area (see story below). Initially, they broke into two groups of about 300 men each. Over time, they split in small groups following different routes which facilitated the search for food. Nothing further was heard of them. This implies that some settled down in various places while others might have fallen victim of the local population.

After King João III was informed through Nuno da Cunha's report of the sinking of the *Conceição* and the *São Sebastião* and the attempt by the men to reach the east coast of Madagascar, he sent the *São Dinis* under the command of Duarte da Fonseca and the lateen caravel *Santa Maria* under the command of his brother Diogo da Fonseca to try to rescue them. Both ships reached the south of Madagascar in late 1530 and anchored in the Bay of St. Luce. While going ashore, a barge in which Duarte da Fonseca, ten soldiers and some crew had climbed capsized and all died in the surf. Captain Diogo da Fonseca took command and continued to look for the survivors along the south coast. Seeing large columns of smoke, he anchored in Ranofotsy cove[38] in early 1531. He found four of the survivors; three of the *Conceição* and one from the *São Sebastião* together with a French sailor who had been left behind by a ship from Dieppe that had passed there in 1527. The sailors told Diogo da Fonseca that many members of the crew had stayed behind in villages during the expedition to the east but they were so scattered inland that it would be impossible to go and fetch them. A discouraged Diogo da Fonseca brought the five survivors to Mozambique. He left for India in April 1531 but both ships wrecked on the island of Socotra (off Yemen).

SÃO ILDEFONSO[39] (1527)

While the 1527 armada was sailing to India, King João III's orders sent earlier concerning the succession of the Governorship in Goa after the death of Don Henrique was a cause of controversy between the designated successor Pêro Mascarenhas, who was in Malacca at the time, and Lopo Vaz de Sampayo, who was named interim Governor. When the King realized that Pêro Mascarenhas' nomination was in jeopardy, he sent a light messenger ship, the *São Ildefonso* commanded by Captain Pêro Aenes Francês (a Frenchman) with Pêro Vaz o Roxo as pilot later in 1527. Their orders were to arrive in Goa before the 1527 armada started to head back to Portugal. They carried a letter with new succession orders stating explicitly that Pêro Mascarenhas was to govern India. Such trips represented the 'express mail' of the time. They used fast ships able to enter the coast easily. They were commanded by Captains with a great deal of knowledge of the sea and the route mixed with a certain amount of madness. Pêro Aenes Francês was one of those, having already completed two of such trips.

While sailing along the southern coast of Madagascar, Pêro Aenes Francês and Pêro Vaz o Roxo decided to anchor in order to pillage, against the King's orders. This disobedience became fatal as the *São Ildefonso* was surprised by a massive storm and foundered.[40] As a result the King's order never arrived. On his return to Goa, Pêro Mascarenhas was not allowed to disembark and went back to Portugal. Many men returned with him to show their support. Eventually, King João III ordered Lopo Vaz de Sampayo's arrest by Nuno da Cunha who was to become the next Governor of India.

Some of the surviving crew proceeded to the Bay of St. Luce and settled there[41] while 76 other survivors reached Ranofotsy cove (also called Galion's Bay), where they settled in a stone house on a small island at the mouth of the Fanjahira river.[42] They were joined later by some of the survivors of the *Conceição* and *São Sebastião* who had walked overland from St. Augustine. The local leaders persuaded the Portuguese to organize a housewarming party which is a customary feast called *Misanasana*. The population brought honey wine and they all met at a shady place on the banks of the river Imorona. The Malagasy numbered 500 or 600. They asked the Portuguese to display their wares, their gold, silver and other goods saved from the wreckages so that they could rejoice from seeing so many treasures. After displaying the riches and having feasted and drunk a lot of the

Figure 15. Cannons believed to be those of the *São Ildefonso*

Figure 16. Cannons believed to be those of the *São Ildefonso*

Figure 17. Cannons believed to be those of the *São Ildefonso*

honey wine the local leaders gave a signal and the Malagasy attacked the Portuguese killing about 70 of them. Only five, who had stayed behind in the stone house with about 30 servants, were saved. During the ensuing months they organized raids against neighboring villages and burned them to the ground in retaliation for the murder of their comrades. These furious raids compelled the local population to ask for a truce. They offered to provide them with as much food as they wished until a Portuguese ship passed by to bring them home.[43] In 1531, the five survivors were picked up by *São Dinis* and the lateen caravel *Santa Maria* under the command of Diogo da Fonseca.[44]

FLOR DA ROSA[45] (1528)

On April 18, 1528 Nuno da Cunha,[46] freshly named by King João III as the new governor of India, left Lisbon on the *Flor da Rosa* with an armada of eleven ships and about 4,000 crew and passengers including more than 2,500 soldiers. He was accompanied *inter alia* by his brothers Simão da Cunha (*Castello*) and Pêro Vaz da Cunha (*Santa Catherina*), Dom Fernando de Limà (*Santa Maria de Espinheiro*) and António de Saldanha (*Nossa Senhora de Ajuda*). The fleet

Figure 18. View of Fort Dauphin from the Portuguese fort (reproduced with the kind permission of Mr. Tim Healy)

transported 200,000 crusados from Portugal and other countries in gold and silver coins, divided up among all the naús for the necessities in India and to cover costs. On July 6 a storm separated the fleet in the area of Cape of Good Hope. Some of the ships reassembled in Mozambique. The *Flor da Rosa* and five other ships were thrust eastward in the Indian Ocean. They decided to sail to Santa Apolonia (Reunion Island) to take on water and supplies. Another storm south of Madagascar again split the fleet and Nuno da Cunha was pushed to Madagascar with the *Santa Catherina* of Pêro Vaz and the *Santa Maria de Espinheiro* of Dom Fernando de Limà. Being in short supply of fresh water, the three ships tried to anchor at Cape St. Mary, Madagascar's southernmost point but bad weather prevented them to do so.

Pushed further west, the three ships tried to get into St. Augustine Bay. The *Flor da Rosa* hit the sandbanks three leagues away from the bay at the same spot where the *Conceição* and the *São Sebastião* foundered the previous year. Fortunately, they dislodged her and safely anchored in the bay on August 23, 1528[47]. After the ships had anchored, inhabitants from the mainland brought sheep, fowl, green beans and other supplies that they bartered for some metal pieces and other objects of little value. Two days later, the chieftains brought over a ragged Portuguese who was clearly in poor health. He was overjoyed at seeing them to the point of being speechless. When he recovered, he informed Nuno da Cunha that the *Conceição* and *São Sebastião* had wrecked there the year before and the two Captains and most of the crew had gone inland a month and a half ago with the goal of reaching the east after the failed rescue attempt by Antonio de Saldanha, while abandoning him because of his ill health. He added that the locals tormented him for as long as he possessed a few items that they desired but they were welcoming and helpful once he had nothing and was naked like them.

On August 26 in the evening of the third day at anchor, a brutal, violent sea wind suddenly erupted. With the winds blowing from the sea the ships could not set sail. At the same time, the barges that went upriver to explore the surroundings and fetch water were unable to return as the race at the mouth of the river was impassable. The *Flor da Rosa*'s dragged on her anchor and was violently tossed around. Before long, the anchor cable ruptured. Another one was thrown and eventually four others but the linen hawsers were ripped apart as they were damaged by the humidity in the hold. She was thrown on a sandbank three fathoms deep. A few massive breakers broke her and keeled her over. The water immediately reached the steerage level. Though they were close to the coast, Nuno da Cunha kept everyone on deck, on the forecastle and on the quarter-deck the whole night to prevent anyone from dying in the furious race if they attempted to swim to shore.

The two other naús had cables made from coco fiber that were not cut by the coral heads, and survived the squall. During the following

Figure 19. Nuno da Cunha (Source, Grandidier, Tome I, p. 64 bis)

days after the storm subsided, the passengers were divided among the other two ships. The masts and yard, the money chest, most of the valuables, the cannons from the main deck, the forecastle, and the quarter-deck, some of the cargo and the anchors were transferred on the *Santa Maria*. On September 3rd, Nuno da Cunha ordered that the wreck be set on fire and it burned up to the water level. Therefore, a large part of the cargo, belonging either to the King or to private passengers was lost. This included cannons, a bronze basilisk cannon[48] which Nuno da Cunha sorely missed as well as all the light arms the men could have made good use of. The two remaining ships then sailed to Malindi via the Comoros and Zanzibar and later seized Mombassa.

SÃO PAULO[49] (1538)

The main armada of naús for 1537, under the command of Dom Pêro da Silva da Gama, left Lisbon on March 12. In May of that year, another fleet of six caravelas departed from Lisbon with Diogo Lopes de Sousa on the *São Paulo* as capitão mor (commander). On her return from India, the *São Paulo* foundered on Madagascar's east coast in the first days of December 1538.

ESPERANÇA (GALEGA)[50] (1540)

Pêro Lopes da Sousa was known as a crafty and experienced sailor and had previously commanded armadas to Brazil. In 1539, he was trusted with command of the armada of the Carreira da India and captained the naú *Esperança* (also called *Galega*). The fleet left Lisbon on March 24, 1539 and arrived in Goa in September. He was faultless in the preparation of the returning ships unlike many other commanders of the Carreira. He was so severe that sometimes he threw goods and even slaves overboard to ensure that his orders were followed. This caused him to be hated by his men. The *Esperança* left India in early 1540 and wrecked at the end of February 1540 in the north of Madagascar on her way to Mozambique with all hands lost.

In 1543, Martim Alfonso da Sousa, the Portuguese governor of India from 1542 through 1545, was worried about the fate of his brother who reportedly had wrecked in Madagascar. He sent one of his Captains, Diogo Soares a half-pirate, to look for him. Captain

Soares anchored in the bay that still bears his name. He came back to Cochin the same year without any information but with spoils from his plunders including a load of silver and slaves.[51]

NOSSA SENHORA DA BARCA[52] (1559)

A fleet of six ships which had sailed from Portugal in 1557, left Cochin in India in January 1559 for the return to Portugal, including the *Nossa Senhora da Barca,* commanded by Don Luis Fernandes de Vasconcelos. In the night of March 17, 1559, she was surprised by a cyclone when sailing about 12 to 15 leagues from Madagascar's west coast near latitude 22° south (opposite Tsingilofilo Bay) and quickly took on water. When the water level in the hold rose to about 6 feet, the ship became hard to navigate. Nonetheless, the pilot tried to steer the ship towards the coast against the furious elements.

Unbeknownst to the pilot and the crew, the officers convinced Don Luis that the *Nossa Senhora da Barca* would sink before reaching shore and the only way to save their lives was to reach the shore in the yawl. Don Luis concurred, had the yawl put to sea and rigged and furtively embarked on the yawl with the officers, six sailors, a barrel of water and some dry biscuits. While the *Nossa Senhora da Barca* continued to sink Don Luis kept the yawl at a distance to prevent crew members who were jumping ship from climbing on board and capsizing the yawl; instead, he shamefacedly picked and chose the people who would accompany him.

When the yawl had 60 passengers, his officers told him that the yawl's capacity was reached and he ordered them to set sail. Noticing that Friar Fernando de Castro had remained on board of the *Nossa Senhora da Barca,* he approached her to take him on board indicating that he would not leave without him. Friar Fernando de Castro replied that he was hearing the confession of the crew and was more concerned with saving the souls of the more than 200 men still on board than saving his own life. Then Don Luis and the crew aboard the yawl asked the friar to pray for them and set sail abandoning the rest of the crew. A short while later they saw the *Nossa Senhora da Barca* engulfed by waves and disappear below the surface.

The next day they reached the coast opposite the Bay of São Iago (either Fiheranana just north of Tulear or St. Augustine) and decided to proceed to the south. They rounded the south of Madagascar and reached the eastern coast living off the meager resources they brought with them. They visited several ports along the east coast where the local population brought them food. In between ports, they collected shells and fished along the beaches where they happened to anchor. Continuing their voyage to the north, they eventually arrived in the bay of Vohemar where they encountered a galliot that was forced by adverse winds to seek relief on its route from India to Mozambique. The Captain, recognizing Don Luis took him and the sixty survivors on board. They wintered in the bay of Vohemar and reached Mozambique in the spring of 1560. Don Luis then proceeded to Goa with a fleet of ships that had left Lisbon in April 1560.

NOSSA SENHORA DA ESTRELA[53] (1660)

The *Nossa Senhora da Estrela,* captained by Manuel Botelho do Amaral, was a naú of 340 tons, armed with four iron and two bronze cannons and manned by a crew of 125. Having loaded a cargo of 30,000 Pieces of Eight (810 kilograms of silver) needed to purchase pepper and other spices in India, 24 cases of coral, four iron cannons needed in India, cannonballs, iron, tobacco, 140 casks of wine and some other miscellaneous merchandise and trinkets, she weighed anchor from Lisbon, bound for India, on April 25, 1660, when Portugal was under 60 years of Spanish rule. She was accompanied by the 1700-ton galleon *Nossa Senhora de Conception,* which was sailing to Mozambique via Brazil, and the 800-ton *Santo Sacramento,* which was bound for Goa. The latter got separated from the *Nossa Senhora da Estrela* already in July when sailing in the South Atlantic.

For some unexplained reason, the *Nossa Senhora da Estrela* wrecked on September 13, 1660, about 7 miles north of Tulear.[54] Twenty-two crew members perished in the wreckage. The remaining 103 officers and sailors managed to reach shore with the help of the longboat. In the ensuing days the Captain organized the salvage of the most precious goods. He then departed for Mozambique with some officers and sailors on the longboat. The lore has it that he took all of the silver with him; not certain, though, whether he could have steered the longboat with such a weight in silver on board.

Eighty nine members of the crew stayed in Madagascar. Most quickly dispersed in neighboring villages. Nine survivors were brought to Surat in March 1661 by the British vessel *HMS Eagle.*[55] A Dutch vessel found another survivor, Antony Fery, in the Bay of St. Augustine in 1663, and brought him to Batavia.

NOSSA SENHORA DO MONTE DO CARMO[56] (1774)

The most recent Portuguese ship recorded to have wrecked off Madagascar was not a naú, but a Royal Portuguese Navy ship, the *Nossa Senhora do Monte do Carmo*. Built in the shipyards in Salvador de Bahia, she was launched on February 7, 1760. After sailing in the Atlantic for more than a decade she was sent on her second voyage to India officially to deliver artillery to Goa. Under the command of Captain-lieutenant Hermogénio de Sousa de Campelo, she left Lisbon on April 23, 1774 accompanied by a merchant ship, the *San Francisco de Paula Santa Eulalia e Almas,* bound for Macao. The *Nossa Senhora* first called at Rio de Janeiro to offload 340 all volunteer conscripts. She departed on June 10 and reached the west coast of Madagascar in early August 1774.

In the evening of August 7, the first pilot ordered reefs taken in the topsail and to furl the mainsail as he expected to find land very soon. He also asked the third pilot to keep close watch during the night and to wake him up if the wind freshened. When the report came at four in the morning the first pilot realized that the mainsail was tacked. He

Figure 20. Picture of the *Nossa Senhora do Monte do Carmo* (reproduced with the kind permission of Dr. Robert Sténuit)

was told that it was deployed because the wind had calmed down and the Captain was present when it was raised. As it was close to dawn, the first pilot left things as they were hoping for an open sea until daybreak. However, at 5:30 in the morning on August 8 the *Nossa Senhora* hit the reef of Salara, broke up, filled up with water and sunk to the level of the main hatch.

The crew launched the yawl and a smaller rowboat. As all the passengers were unable to fit in these two crafts, they built a raft with spare yard and small masts that were on board; a difficult task as the race was breaking close to the wreck. They tried to reach the close-by land during the day but could not pass the reef running parallel to the coast. This compelled them to go seaward away from the reef. At nightfall they found a pass allowing them to reach shore. After spending the night on the beach they walked south until they found a village. The local population provided them with supplies in exchange for some of their clothing and the few items that they had salvaged from the wreck. The Captain, the first pilot and sixteen sailors then sailed with the lifeboat to fetch help in Mozambique. They came back a month later with the needed amenities to rescue the crew but many sailors had already died from tropical diseases.

While in Mozambique, the Captain met a lord, His Excellency Senhor Jose Pedro da Camara who was en route from Rio de Janeiro to Goa. He deplored the loss of the *Nossa Senhora*, and the various Petrexos (commodities) that she was carrying and that were needed in Goa. Implying by that that she was carrying more than just artillery. What this extra cargo could be was the riddle that the Belgian archaeologist Robert Sténuit, who discovered the wreck in 1986, wished to crack. As he recounts,[57] the first sight was a field of about 60 neatly aligned cannons as they must have been placed at the bottom of the hold.

After breaking up, fragments of the *Nossa Senhora* were washed in a perfectly straight northeastern line including some of the ship's cannons, two anchors now lying on top of each other and two other anchors further up. Robert Sténuit found some 6,400 reis gold coins with the effigy of King Dom José (who reigned from 1750 to 1777), dating from the late 1760s and minted in Rio de Janeiro and lots of glass shreds indicating that the *Nossa Senhora* was bringing to India a load of rectangular green glass panels encased in lead for assembly into stained glass windows.

But what else could there be? The glass panels must have been loaded in Lisbon together with other European consumables and lots of trinkets that officers and passengers used to stuff their cabins with on top of their statutory caixas de libredad (bauble trunks). These sold for twenty times their purchase price in India. Reviewing notes on Luso-Brazilian trade at the time, Robert Sténuit deduced that the *Nossa Senhora* must have transported, beyond the artillery, bags of sugar, bales of tobacco, and skins and hides which disintegrated in the water. No diamonds or gold (beyond the few coins), but lots of trinkets, including small glass bottles, colored glass pearls, copper bracelets, tin buttons, shoe buckles, flint stones, pieces of ceramic and lots of pink coral beads from the Mediterranean.[58]

REFERENCES

[12] These introductory paragraphs are largely inspired by Gilbert, William, op. cit.

[13] Gilbert, W., op. cit, p. 2.

[14] Guinote, Paulo J.A., "Ascensão e Declinio da Carreira da India (Seculos XV-XIII)," at http://nautarch.tamu.edu/SHIPLAB/01guifrulopes/Pguinote-nauparis.htm. In the first half of the 16th century, the number of ships averaged 8 to 10 per year, falling to 5-6 ships per year thereafter.

Figure 21. Part of the aligned cannons of the *Nossa Senhora*

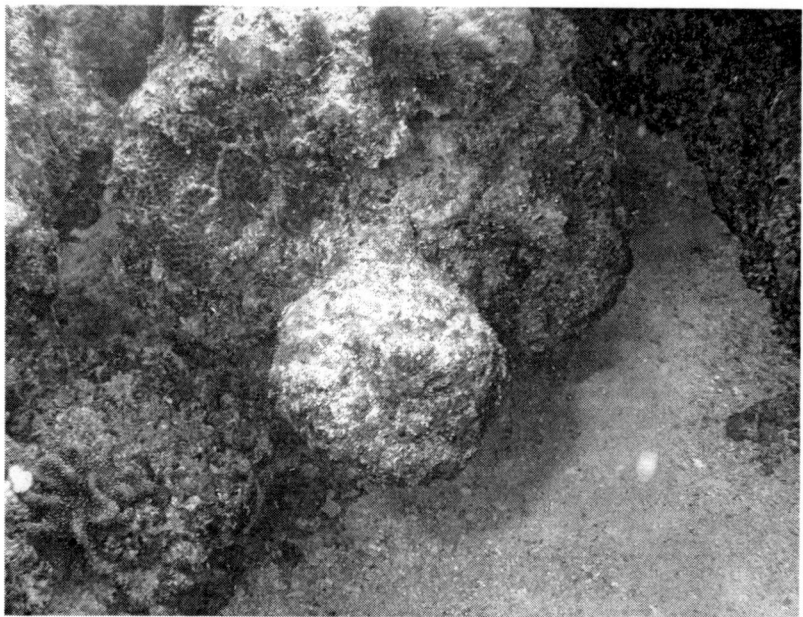

Figure 22. Cannon knob and neck

Figure 23. Crab having found a house in a cannon muzzle

Figure 24. Standalone cannons

Figure 25. Anchors lying on top of each other

Figure 26. Main anchor

Figure 27. Another cannon

Figure 28. Pink coral beads

[15] There is reasonably good documentation on the composition of the fleets of Portuguese East Indiamen that plowed the maritime route to India almost every year from 1498 to about 1650, as well as their routes, their ports of call, the names of the ship's Captains, the Asian governors, and the notable soldiers. Existing knowledge of the naús themselves, however, is scant, being based on a handful of texts and treatises, a small number of representations in charts, drawings, and paintings, and around twenty shipwrecks discovered so far (see Filipe Castro, 2006, India Route Ships Project: Introduction, World Wide Web, URL, http://nautarch.tamu.edu/shiplab/, Nautical Archaeology Program, Texas A&M University).

[16] Before long the local name of Madagascar became accepted as its designation and was indicated on maps as such.

[17] Gaspar Correa in "Historia do descobrimento e primeira conquistas da India" (1516) clearly describes this first discovery of Madagascar. Some other 16th century authors place the discovery in 1506 by other navigators. See Grandidier et al., op. cit., Vol. I, p. 5.

[18] Guinote, P., op. cit., pp. 20-24. This is equivalent to about 20 percent the total number of trips. There is a relatively even distribution over time. However, during the years 1497-1550, about 60 percent of the ships were lost on the outward voyage, due presumably to the lesser quality of the ships and knowledge of the routes and maritime conditions. The reverse is true during 1551-1600: about 70 percent of the ships were lost during the return voyage, often due to

overloading that reduced seaworthiness. For a list of Portuguese Indiamen shipwrecks, see Guinote, P., Frutuoso, E., and Lopes, A., "Naufragios e Outras Perdas da Carreira da India Portuguesa," Lisboa: Ed. Grupo de Trabalho do Ministério da Educação para as Comemorações dos Descobrimentos, 1998.

[19] Guinote, P., op. cit., p. 23. This percentage drops over time, from 38.8 percent in 1497-1550, to 20 percent in 1551-1600, and to 18.9 percent in 1601-1650.

[20] These events have been related by Correia, G., "As Lendas da India,"1555 (and reprint Porto: Lello e Irmão, 1975), t. I, p. 662 and pp. 665-668; de Albuquerque, F., "Commentarios do Grande Alfonso de Albuquerque," 1557, Parte 1, Ch. VIII, IX and X, and 1776 edition, pp. 33-43; João de Barros, "Da Asia Portuguesa," 1778 edition, Dec. II, liv. 1, ch. 1, pp. 7-18 and ch. VI, pp. 87-88; Lopes de Castanheda, F., "Historia de los descobrimentos e Conquista da India pelos Portugueses," 1552 (1833 edition), t. II, ch. XXX and XXXI, pp.101-124; Manoel de Faria y Sousa, "Asia portuguesa," 1666, t. I, p. 95; Osorio, J., "De rebus Emmanuelis,' (1574), translated in 1804, "Da vida e feitas del Rei Dom Manoel," XII livros, t. II, liv. V, pp. 21-26; Maffei, P., "Historiarum Indicarum," Lyon, 1637, Libri XVI, liv. 1, p. 35 and liv. III, p. 121; and Lafitau, P., "Histoire des découvertes et conquêtes des Portugais dans le Nouveau Monde," 1733, t. 1, p.254-255; and Grandidier et al., op. cit., Vol. I, pp. 14-43.

[21] The Primeiras and Segundas Archipelagoes extend from the Mozambique coast off Angoche, in the Nampula Province, to Pebane, on the coast of Zambezia.

[22] Now called Ambinany Matitanana, along the banks of the river of the same name, in the southeast of Madagascar.

[23] Correia, Castanheda and Lafitau, op. cit., claim that Captain Rui Perreira first visited several ports, on the western coast of Madagascar, including the Boina Bay, which he reportedly called Bahia formosa.

[24] Coastal inhabitants in the South of Madagascar have a longstanding tradition of wearing silver jewelry and using silver to line funeral coffins. As no silver is mined in Madagascar, it is thought that the first silver jewelry came with Arab merchants, who were known to trade gold and silver for slaves and rice. The main supply source, however, was silver coins recovered from shipwrecks, that were then melted to make jewelry.

[25] The *São Iago* being deemed too big for this expedition, Tristan da Cunha boarded the *São Antonio* commanded by Captain João da Vega. The other ships included the S*ão Vincente* (Captain Rui Perreira de Coutinho), the *Nossa Senhora da Luz* (Captain João Gomes de Abreu), the *Cirne* (Captain Alfonso de Albuquerque), the *Santo Espirito* (Captain Antonin do Campo), the *Rei Pequeno* (Captain Manoel Teles), the *Rei Grande* (Captain Francisco de Tavora), and ships commanded by Job Queimado and Tristan Alvares.

[26] Tristan da Cunha named it the Bay of Conception (Bahia de Maria de la Conceição), as he landed there on December 8, the day that the Church is celebrating that feast.

[27] Now called Anorontsanga in Rafala Bay, south of Nosy Be.

28 Now called Amber Cape, or Tanjon'ny Bobaomby in Malagasy.
29 This story is recounted by Lopes de Castanheda and Osorjo, op. cit, loc. cit.
30 The *Nossa Senhora da Luz* shipwrecked in 1507 off Pate Island in Kenya.
31 Ranofotsi Cove in the south of Madagascar, a little west of Taolagnaro (Fort Dauphin). This is related by João de Barros, op. cit., Dec. II, liv. IV, ch. III, pp. 391-395.
32 Sources: de Barros, op. cit., Dec. III, Livro I, Ch. II, pp. 14-15; Castanheda, op. cit., Ch. XXV, p. 908; Correia, op. cit., Vol. II, Ch. VI, p. 484 and ibid., "Crónicas de Don Manuel e de Don João III," Academia das Ciências, Lisboa, 1992, p. 113; da Costa Quintella, I., "Annaes de Marinha Portuguesa," 1839, T. I, pp. 326-327; da Fonseca, Q., "Os Portugueses no mar: Memórias Históricas e arqueológicas das Naus de Portugal," 1983, Vol. 1, Lisboa: Associação dos Arqueólogos Portugueses, pp. 223; Xavier, Padre Manuel, "Relações da Carreira da Índia" Dirigido e comentado por Luís de Albuquerque; Lisboa: Alfa, 1990, p. 112; Guinote, P., Frutuoso, E., and Lopes, A., "Naufragios e Outras Perdas da Carreira da India: Séculos XVI e XVII" Lisboa: Grupo de Trabalho do Ministério da Educação para as Comemorações dos Descobrimentos Portugueses, 1998, p. 197.
33 Today called Kilwa Kisiwani, an island off the coast of present day Tanzania which the Portuguese took control over in 1505.
34 The story is recounted by Correia, op. cit., Vol. III, Liv. IV, Ch. XVI, pp. 225-230; Diogo do Couto, "Décadas da Asia," Dec. IV, liv. III, ch. V, pp. 206-207, and liv. V, ch. I, pp. 331-339; de Barros, J., op. cit., Dec. IV, Liv. III, Ch. I, II, and III, pp. 256-261; Grandidier et al., op. cit., T. I, p. 58-59; Xavier, Padre Manuel, op. cit.; Sousa, Frei Luis de, "Anais de D. João III"; Lisboa; Livraria Sá da Costa, 1938, Livro IV, Ch. 1, pp. 1-2; and Guinote, P., "India Route Project: *Armadas que partiram para a Índia (1509-1640),*" World Wide Web, URL, http://nautarch.tamu.edu/shiplab/, Nautical Archaeology Program, Texas A&M University, 2003.
35 Sometimes confused as the *São António* (see da Fonseca, Q., op. cit., pp. 279 and 295).
36 The *São Sebastião* (sometimes called *Bastiana*) had already sailed to India in 1524 under the command of Captain Pero Mascarenhas, aboard which Vasco de Gamma undertook his third and last voyage.
37 Correia relates that the ships wrecked in the south of Madagascar, with the *Conceição* foundering a little southward of the *São Sebastião*. Do Couto mentions that they wrecked on the southwest coast, in the Bay of São Iago, by 20½° south (identified by Grandidier as Tsingilofilo Bay, to the south of Morombe). Albert Kammerer ("La Découverte de Madagascar par les Portugais et la cartographie de l'île :1500-1667," Boletim da Sociedade de Geografia de Lisboa, Série 67, N° 9-10, 1949, pp. 584-586) correctly points out that there is no bay at or near that latitude. Given the detailed description of the Bay of São Iago given after the wreckage of the *Flor da Rosa* there a year later by de Barros, by far the best historian of the Portuguese discovery, Kammerer deduces that

the *Conceição* and the *São Sebastião* foundered a little south of St. Augustine Bay (possibly at the large reef surrounding the small island of Nosy Ve), by latitude 23°25' south.

[38] Where the Portuguese had built a stone house (see below). Off the town of Italy, west of Fort Dauphin. João de Barros, op. cit., Dec. IV, liv. V, ch. vi, p. 583, nota, and liv. III, ch. ii, p. 261.

[39] Sources: Correia, op. cit., Vol. III, Liv. IV, ch. XVI, p. 225 ; de Barros, J., op. cit., Liv. IV, Vol. 1, pp. 40-41, Castanheda, F. L., op. cit., Vol. II, p. 394 ; Grandidier et al, op. cit., t. I, p. 60; Sousa, Frei Luis de, op. cit., loc. cit. ; Decary, R., "Les voyages des Portugais à Madagascar au 16ème siècle," 1949, p. 192 ; and Cruz, M. A. L., "As viagens extraordinárias pela rota do cabo" in *Actas do VII Seminário Internacional de História Indo Portuguesa;* Angra do Heroísmo; [s.n.], 1998.

[40] Correia, t. III, p. 309, indicates that one of the survivors was picked up by Nuno da Cunha aboard the *Flor da Rosa* in 1528. This is probably incorrect, and a confusion with the survivor of the *Conceição* and *São Sebastião*.

[41] Some of their grandchildren met the survivors of the Dutch ship *Westfriesland* that wrecked there in 1606 (see below).

[42] Anosy island, also called the Island of the Portuguese or Trano vato, meaning house built from stones. Flacourt (1661 edition) and Grandidier (t. VIII, p. 58, note 1) mention that they built the stone house. In his annotated edition of Flacourt, Claude Allibert ("Flacourt, édition de 1661 annotée et présentée par Claude Allibert," Paris: Inalco-Karthala, 1995, pp. 16-17 and 482-483), argues that they occupied a pre-existing structure. A manuscript of Diogo Lopes de Sequeira (who picked up two of the surviving crew of the *Nossa Senhora da Luz* at Ranofotsy Cove in August 1508—see above) dating from 1508 (cited in Axelson, E., "Southeast Africa: 1488-1530," London-New York-Toronto: Longmans and Green, 1940, p. 105) already mentions the stone house, indicating that it was built by previous immigrants, possibly the surviving crew of sailors from Gujarat in Northern India who had wrecked there in the first half of the 14th century. Furthermore, Vérin and Heurtebize, "La trañovato de l'Anosy, première construction érigée par des Européens à Madagascar. Description et problèmes," Taloha 6, 1974, pp. 117-142 (cited in Allibert, op. cit., p. 482), indicate that the stone house was located near a royal Malagasy village, the Kings of Fanshere. The surviving Portuguese crew thus did not chose that spot randomly, but to benefit from the protection of the ZafiRaminia.

[43] The story has been related, with some errors, by Flacourt, E., "Histoire de la Grande Isle Madagascar," 1658 edition, pp. 32-33. However, this massacre has not been confirmed by excavations done by Vérin and Heurtebize (op. cit.).

[44] More than 80 years later, in November 1613, Captain Paulo Rodrigues da Costa of the caravel *Nossa Senhora da Esperanca* and two Jesuit priests, Pedro Freire and Luis Mariano, visited Ranofotsy cove and the stone house, and named it San Lucas. They abducted the king's son, Andrian-dRamaka, and brought him to Goa to live for a couple of years at the court of the viceroy. An-

drian-dRamaka was returned in April 1616 and went back to live with his parents. Four Jesuit priests who were part of the return voyage went to live on the island at Ranofotsy cove with the goal to convert the neighboring populations to the Christian faith. Beset by illness and the failure of their enterprise, they abandoned the mission 11 months later and returned to Goa.

[45] Related by Correia, op. cit., t. III, p. 309; de Barros, J., op. cit., Dec. IV, liv. III, ch. III, pp. 264-270 (of the 1778 edition, and ch. I, pp. 256-259 of the 1945-46 edition); do Couto, op. cit., Dec. IV, liv. V, ch. I, pp. 331-332, and ch. iii, pp. 333-339; de Andrada, F., "Cronica del Rey Don João III," 1613, secunda parte, ch. XLVII, pp. 66-67 (and 1976 ed., ch. XXXXVII, pp. 422-423); and Grandidier et al., op. cit., t. I, pp. 63-76.

[46] Nuno da Cunha was the son of Tristan da Cunha and Dona Antonia d'Albuquerque. He had explored Madagascar with his father aboard the *São Antonio* in 1506 (see above). He was the tenth governor of Portuguese India between 1528 and 1538. He died in 1539, at the age of 52, just after having returned from India.

[47] Do Couto and de Andrada relate, probably erroneously, that Nuna da Cunha anchored in a bay on the east coast, where he rescued a survivor of Pêro Vaz o Roxo's ship.

[48] The largest and heaviest fortress cannon in use at the time, which Nuno da Cunha most probably wanted to install in some fortress in India, most certainly abundantly decorated, dated and signed.

[49] Sources: da Fonseca, Q., op. cit., p. 323; "Introdução a Relação Das Náos e Armadas da India Com os successos dellas que se puderam saber, Para Noticia e instrucção dos curiozos, e amantes Da Historia da India," leitura e anotações de Maria Hermínia Maldonado, Coimbra: Biblioteca da Universidade, 1985 (also at www, url, carreiradaindia.net/index.php?paged=2&s=moor); Xavier, Padre Manoel, op. cit.; Furtoso, E., "India Route Project: *Relação de Capitaens Mores e Naos que Vierão do Reyno a este Estado da India des do seu Descobrimento*," World Wide Web, URL, http://nautarch.tamu.edu/shiplab/, Nautical Archaeology Program, Texas A&M University, 2003; and Guinote, P., op. cit.

[50] Sources: Correia, op. cit., t. III, ch. XXXV; Xavier, Padre Manuel, op. cit., Pissarra, J.V.A., "Navegações Portuguesas, Sousa, Pêro Lopes de," 2002, at WWW, URL, http://www.instituto-camoes.pt/cvc/navegaport/g61.html.

[51] Related by Correia, t. IV, ch. XIII, pp. 266 and 275, and Grandidier et al., t. I, p. 89.

[52] Narrated in Do Couto, Década VII, Livro 5, cap. II, and *ibid.* Livro VIII, cap. 1, pp. 173-179; Bernardo Gomes de Brito, "História Trágico-Maritima" (I, 311-49, of the 1735 edition); Grandidier et al., op. cit, t. I, pp. 109-111; Duffy, J., "Shipwreck and Empire. Being an account of Portuguese maritime disasters in a century of decline," Harvard University Press, London: Geoffrey Cumberlege, 1955, pp. 30-I; Boxer, C.R., "An introduction to the História Trágico-Marítima," (reprint) from the Miscelânea de Estudos em honra do Professor

Hernâni Cidade," Lisbon, 1957, pp. 14-15, and *ibid.* "Further selections from the Tragic History of the Sea, 1559-1565," Cambridge University Press, 1967, p. 53.

[53] Source: de Barros, op. cit., Tomo 6, Parte 2, p. 561.

[54] Six of her cast iron cannons washed on the beach.

[55] "Dagh-register gehouden int Casteel Batavia," Batavia, 1661, June 1 and 21, and July 11, pp. 158, 190, and 212; Grandidier et al., op.cit., vol. III, pp. 295-296.

[56] Source: Sténuit, R., "Ces Mondes Secrets où j'ai plongé," Paris: Robert Laffont, 1988, ch. XXVI, pp. 386-396; ibid., "La Vergine di Monte Carmo," *Mondo Sommerso,* 1991, 352: 54-63; ibid., "Nossa Senhora do Monte do Carmo," *Tauchen,* 1992, 7: 52-58.

[57] Sténuit, R., op. cit., pp. 392-396.

[58] Robert Sténuit found more than 600 of different sizes, and this author, who dove the *Nossa Senhora* on several occasions, also found many of these beads.

2
The Dutch United East India Company

The Dutch, a traditional trading and seafaring nation, joined the race for the trade in the East as their political independence grew. They eventually became the Portuguese biggest competitors on the East India run and for domination of the spice trade.[59] Starting in the 1550s, Portugal's economy declined, wounded by the religious zealotry of the Counter Reformation. Its empire was fading and it could ill afford the cost of administering its colonies and running trading posts abroad—including a constant need for fresh supplies of weapons, ships and men. A defeat at the hands of its Muslim foes, the Moors, in 1578 further weakened the Portuguese empire. Its independence disappeared when it fell under Spanish rule in 1580.

After Charles V inherited the Low Countries and Spain in the early 16th century, he disregarded the charters giving rights of freedom to many of the large cities of the Low Countries. Needing money for his endless wars, he levied heavy taxes on the towns and cloth industry. Furthermore, when Calvinism spread he introduced the Inquisition and tried to root it out. Philip II, a devout Catholic, took over Spain and the Netherlands from his father in 1556. His repression of the Protestants supported by the zeal of the Jesuits and the Spanish inquisition lead the Dutch to openly threaten Spanish rule. They rebelled in 1566 and by 1577, Dutch Protestants led by William of Orange governed virtually all of the Netherlands, with the Dutch United Provinces declaring their independence from Spain in 1579.

This was too much for Philip II. He sent his nephew Alexander Farnese, Duke of Parma, into the country with a powerful army to put down the rebellion. The war lasted forty years with varied fortunes. Parma's "Army of Flanders" started capturing the rebel towns one by one. The southern provinces (Belgium today) were soon subdued and accepted firm Spanish rule. The Prince of Orange was assassinated in 1584 but the struggle went on under his second son, Prince Maurice, a boy of seventeen. By 1585, Parma had captured the key port of Antwerp and driven the Dutch rebels back to their northern strongholds. Queen Elizabeth sent English troops to rescue them from collapse. Finally, in 1609 a truce was established, but Spain did not recognize the Dutch Republic until 1648 as one of the provisions of the Treaty of Westphalia.

The war of independence from Spain energized the Northern provinces of the Netherlands. They also benefited from a large number of emigrants and enterprising merchants from the southern provinces. They came chiefly from Antwerp, a town which had for many years enjoyed considerable, though indirect, share in the trans-Atlantic trade of Spain and Portugal. These men were animated by the bitter hatred of exiles enhanced by difference of faith and the memory of many wrongs. The idea that arose among them was to deprive Spain of her commerce, cripple her resources and strengthen those of the Protestants and eventually to force the southern provinces of the Netherlands from their oppressors.[60] When the Spaniards forbade Dutch vessels to carry on any traffic with Spain and took away their share in trans-Atlantic commerce the Dutch determined to find a way to Asia. In the same vein, the Dutch felt that the capture of Portuguese outposts and trade in Asia would be an important second front that would lead to the eventual defeat of Spain.

The Dutch expansion to Asia started as a series of expeditions financed by several trading companies that were set up in various ports in the Dutch Republic with the goal of trading for pepper and spices. However, the intense rivalry between these companies resulted in a steep fall in the price of spices sold in Holland. In 1601, the *raadspensionaris* of the Dutch States General, Johan van Oldebarnevelt, launched the idea of merging all the existing companies into one United East India Company (*Verenigde Oost-Indische*

Compagnie or VOC). On March 20, 1602 this enterprise was inaugurated as a joint-stock company with a founding capital of 6.5 million Dutch guilders and granted monopoly on the trade "East of the Cape of Good Hope."

Beyond its trading venture—thought by many to be the world's first ever multinational company[61]—the VOC was also designed to play a role in the Dutch struggle against the Spanish crown. The States General of the Dutch Republic provided the Company's ships with the guns and ammunition needed to attack the Spanish and Portuguese colonies in Asia, and some of the VOC's profits flowed back to the State. The VOC made a vital contribution to the successful end of the eighty year war of independence (1568-1648) against the Spanish crown.

Of some 2,000 ships that were sent to the East by the VOC, about one third (653) were lost. About eight are reported to have wrecked

Figure 29. Main Dutch getaways

Figure 30. A typical Spiegelretourship (authentic reconstruction of the VOC ship Batavia from 1628. Built by and to be seen daily at the Bataviawerf in Lelystad, Holland, www.bataviawerf.nl; Copyright, Bataviawerf 2007)

off Madagascar. All of these wreckages took place on Madagascar's south or east coast as the Dutch ships generally avoided the Mozambique Channel. Initially, the VOC fleets sailed towards Mauritius which the Dutch had taken over.[62] By 1617, it was mandatory for VOC fleets to call at the Dutch settlement in Tafelbaai, off the Cape of Good Hope, for fresh supplies. From there they could sail non-stop to India via a new route which shortened the trip by about three months: instead of crossing the Indian Ocean on a northeastern course, they went a little south to the 35th parallel, then went on an eastern route for about a thousand miles before switching to a northern course to reach the Sunda Straight (which runs between Java and Sumatra).

The Dutch seafarers learned Madagascar's east coast well.[63] The first expedition of the Old Company (*Oude Oost Indische compagnie*), financed by the Compagnie van Verre of Amsterdam, consisted of three well-supplied and armed ships, the *Mauritius,* the *Hollandia,* and the *Amsterdam,* together with a small pinas, the *Duyfken.* It left Texel in April 1595 under the command of Cornelis de Houtman. After passing the Cape, in early September they decided to halt in

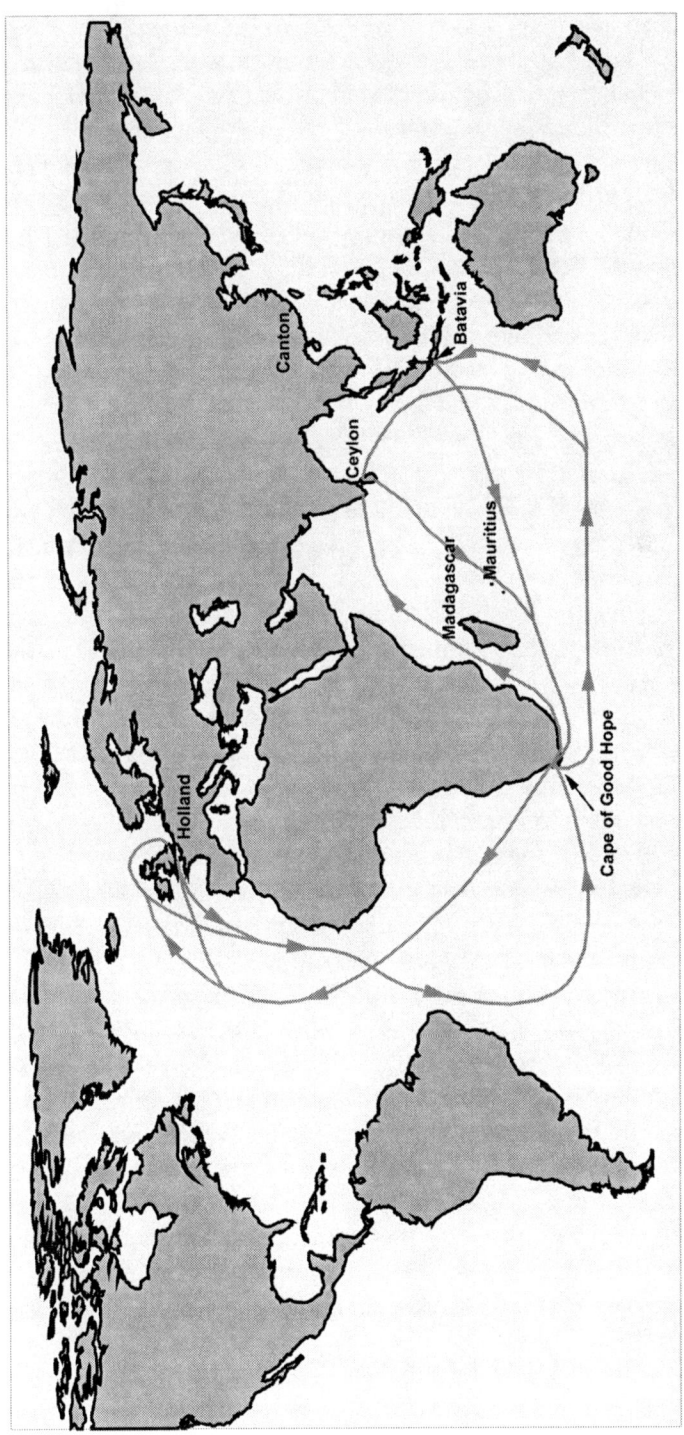

Figure 31. Dutch sea routes to and from Asia

Madagascar as many sailors were suffering from scurvy. Adverse winds and currents prevented them from passing the southeastern point of Madagascar. They anchored in the Bay of Ampalaza (the southwestern point of Madagascar) in mid-September 1595 and stayed there for three weeks. They buried 28 sailors who died on a small island at the entrance of the bay (Nosy Manitsa) which they named "Dutch cemetery." Then they sailed north to the Bay of St. Augustine (off Tulear) where they wintered for two months. In mid-December 1595, they weighed anchor and rounded the south of Madagascar, reached the island of St. Marie off the east coast in January 1596 before calling at the Bay of Antongil. From there they left for India on February 12, 1596.

The second expedition of the Old Company, consisting of eight ships, left Texel on May 1, 1598 under the command of Admiral Jacob van Neck. Shortly after rounding the Cape of Good Hope on July 28, 1598 they were caught by a major storm and separated. Five ships, the *Amsterdam, Utrecht, Zeeland, Gelderland,* and the yacht *Friesland,* put in on August 24 at the point of Fenambosi in the southeast of Madagascar, their first landing since leaving Texel. From there they proceeded to Mauritius and claimed it for Holland before moving on to Bantam, a city and former sultanate on Java in Indonesia. The other three ships, the *Mauritius, Hollandia,* and the small yacht *Overijssel,* put in at the Island of St. Marie and the Bay of Antongil before proceeding to Bantam. At St. Marie, they watched the extraordinary sight of local fishermen hunting and catching whales.[64]

The fourth expedition commissioned by the Amsterdam chamber of the Old Company, consisted of seven ships under the command of Admiral Steven van der Haghen aboard the *Zon*. They left Texel on April 6, 1599, bound for Bantam. Unable to pass the southeastern point of Madagascar because of adverse currents they put in on October 27, 1599 into a bay which they called Sonnebaai (Sun bay, north of Andovoranto). The fleet left the bay on November 10 but the winds died and currents pushed the ships back to shore. One of the ships, the 500-ton yacht *Wassende Maan,* under the command of vice-admiral Kornelis Heinszoon, nearly ran aground on the rocks. Eventually the fleet rounded the south of Madagascar and sailed north along the east coast. From November 17 till December 21, 1599 they had to lie in the Bay of Antongil waiting for favorable winds to sail on.[65]

Many other Dutch fleets put in at Madagascar for refreshments over the years with the Bays of St. Luce and Antongil being the spots

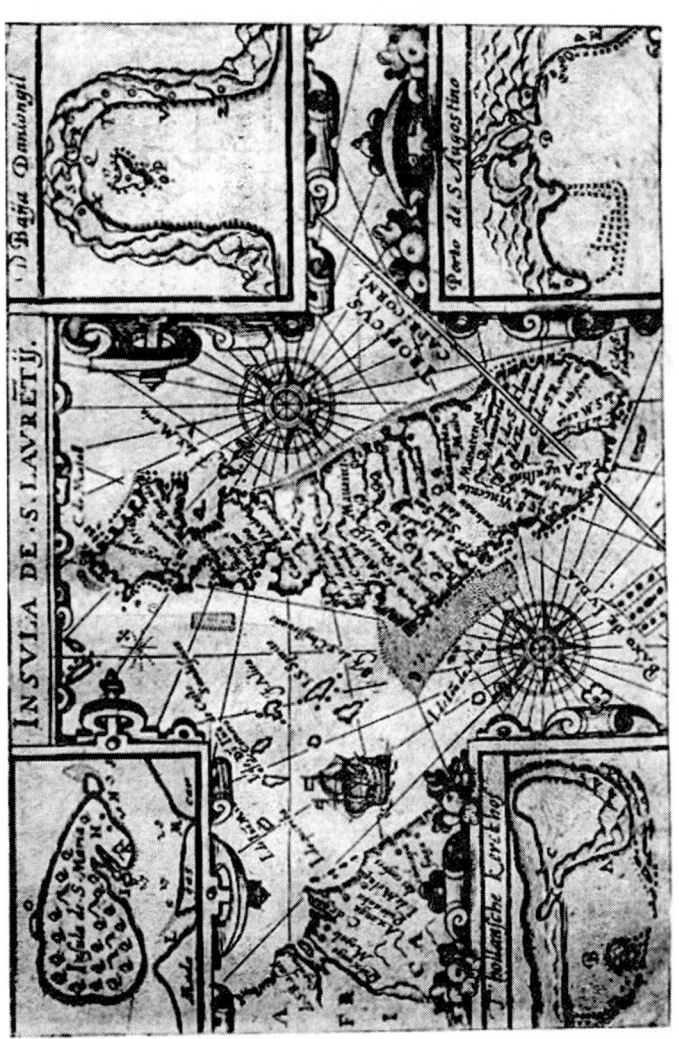

Figure 32. Map of Madagascar by Houtman (1595) with details of the Dutch cemetery on Nosy Manitsa, the Bay of St. Augustine, the Island of St. Marie, and the Bay of Antongil (source: Grandidier, op. cit., T. 1, p. 170 bis)

Figure 33. Local fishermen catching whales off the Island of St. Marie (source: Grandidier, op. cit., T. 1, p. 246 bis)

of choice. In 1642, the Dutch set up a trading post (factorij) in Antongil to provide slaves for the settlement on Mauritius. It was closed four years later as many of the Dutch died from tropical diseases and the survivors clashed with the local king.

ALKMAAR[66] (1604)

The *Alkmaar* belonged to the so-named Atjehse fleet. Their destination was the now-named Aceh on the island of Sumatra in Indonesia. The fleet consisted of eight ships, *Amsterdam, Alkmaar, Hoorn, Enkhuizen (Bruinvis), Zwarte Leeuw, Witte Leeuw, Rode Leeuw* and *Groene Leeuw*. This fleet was part of the so-called "Vijfde Shipvaart" (fifth ship sailing) that also included the Molukse fleet. Alkmaar sailed her maiden voyage under the command of Captain Pieter Stokmans out of Texel on April 23, 1601 and arrived in Bantam on February 22, 1602. She left Bantam for the return voyage on October 10, 1603. She was affected by shipworm and left behind in the Bay of

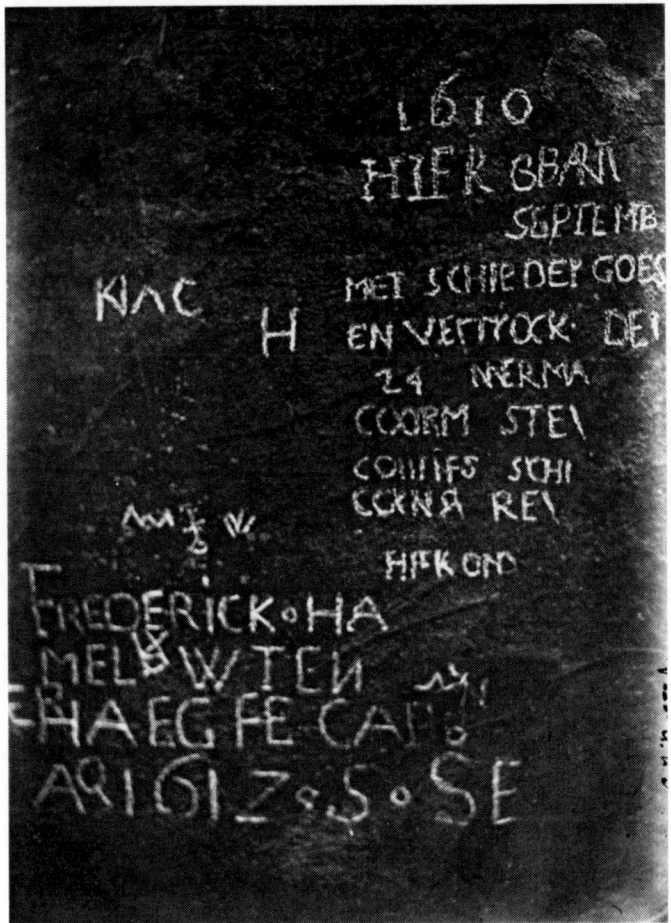

Figure 34. Rupestrian writings left by passing Dutch sailors in the 1600s on the rocks of Nosy Mangabe at the Bay of the Dutch in Antongil (Photography by E.Drouhard, 1927, reproduced with the kind permission of the Bibliothèque Grandidier, Antananarivo, Madagascar)

Antongil in July 1604. The *Hof van Holland* was sent from Amsterdam to pick up her cargo and returned to Texel in April 1605.

WESTFRIESLAND (1606)

The 700-ton *Westfriesland*, built in Enkhuizen and mounting 50 guns, was commissioned by the Enkhuizen chapter of the VOC. She left for her maiden voyage under the command of Captain Jakob Jakobszoon Klunt from Texel on December 18, 1603 bound for Ban-

Figure 35. Rupestrian writings (continued)

tam, which she reached on December 31, 1604. *Westfriesland* left Bantam on August 25, 1605 for the return voyage, under the command of Willem Janszoon van Amsterdam and loaded with spices. Taken in a storm she ran ashore in the bay of St. Luce (Manafialy) in 1606.

While the crew was cutting wood to build a vessel to return to Bantam some local inhabitants came to meet them. Believing they were Portuguese, they told them in Portuguese that they were the grandchildren of their countrymen and asked whether there were some priests on board.[67] Learning that the crew was Dutch, they told them that many decades ago another big ship was lost and that the crew had fled inland, eventually occupied this part of the island where they had

Figure 36. Rupestrian writings (continued)

Figure 37. Rupestrian writings (continued)

married local women and had many children. As their forefathers always wished to have priests instruct them they now had the same wish. They showed the Dutch sailors the tomb of their first king, the Captain of the lost ship, which was adorned by a beautiful cross and a tombstone with worn out and damaged inscriptions which were impossible to read.

After the crew made it back to Bantam, they told of this to the Portuguese Augustine friar Athanase de Jesus, who was their prisoner. He related it to the Archbishop in Goa.[68] This bishop advised Jesuit missionaries who were leaving for Mozambique to try to obtain more information on these people and to help them if needed.

BRAK[69] (1613)

The 100-ton yacht *Brak* was commissioned by the VOC in 1609. She left Texel on her maiden voyage under the command of Jakob IJsbrandszoon[70] on January 30, 1610 hired by the trading chamber Enkhuizen. She was destined for Bantam and a more extensive discovery of New Guinea. The ship was part of a fleet led by Governor-General Pieter Both. The fleet consisted of the *Wapen van Amsterdam, Brak, Ceylon, Witte Leeuw,* and *Zwarte Leeuw*. Three other ships coming from Zeeland (*Ter Goes, Oranje* and *Vlissingen*) met the main fleet on February 1, 1610 near the island of Wright. On June 9, the fleet was scattered in a storm and the *Brak* lost contact with the fleet on June 23. On August 3, 1610 she reached Mauritius. She then sailed, under the command of Jacob Franszoon, via Hirado, Patani and Johor to Bantam which she reached in early 1612. During her return voyage, *Brak* was taken in a storm off the southern coast of Madagascar. Overloaded she soon started listing and taking in water. A few hours later she sailed aground with all but one hand lost on the coast of Karimboly on August 7, 1613.

The survivor was the son of the Captain, a twenty-year old named Pieter. He was lucky enough to take hold of an empty cask that was floating among the debris. He swam for three days holding on to the cask, with the waves pushing him closer to the coast each day. A wave finally washed him ashore in the midst of some local inhabitants who were picking up debris from the wrecked ship. Seeing this blond lad emerging from the sea they wondered whether they were seeing a ghost or some supernatural being. Pieter was so feeble that he could hardly lift his body. Seeing the locals surround him he cupped his hands to ask for something to drink. The locals were puzzled by the gun he carried on one side and his sword on the other side. What should they do with this foreigner, fell him on the spot or worship him? In doubt they sent for their chief Andriamamory.

The chief came to the shore, felt pity for this young lad, and had him transported to his village. He put Pieter in his home, gave him a fresh change of clothes, fed him and treated him as well as he could. Pieter had no goods to trade but carried two splendid diamonds set in golden rings that his father had given him at the time of the wreckage, telling him to use them if he were to survive. Pieter gave one of the diamond rings as a present to Andriamamory and kept the other one for himself. Pieter lived with Andriamamory two years. He lived a real Malagasy life and learned the local language. He hunted and fished with the natives and taught the villagers how to make large and solid fishing nets. Deft and courageous he was liked by all.

The king of the Anosy province Andrian-Tsiambany heard that there was a Christian in the Karimboly area. He sent 13 oxen as a gift to Andriamamory and requested that the boy be given to him. So Pieter went to live in Anosy where he was very well received. Andrian-Tsiambany gave him a house, one of his daughters to keep him company and entertain him and some slaves to help him out. He spent five years there perfecting his knowledge of the language. They respected and feared him. He was a bit of a sorcerer. To entertain the company in the evening he put on his ring, covered the fire and the diamond was so bright that it lit up the whole cabin. The Malagasy silently admired but also dreaded the brilliance of this stone.

Another Dutch ship came to anchor in the Bay of St. Luce for water. Andrian-Tsiambany sent Pieter to greet the Captain and bring him welcome presents giving Pieter presents to distribute to his fellow countrymen. The Dutch were surprised to hear him speak their language as they believed that he was the king's son until he told them who he was. The Captain, charmed by the encounter and the presents, gave 100 Pieces of Eight, linen from India and some porcelain to bring to the king and gave another half of it to Pieter for himself. Pieter escorted the Dutch officers on their way to the court to thank Andrian-Tsiambany for all the good things he had done for their countryman during these years. Growing home-sick, Pieter returned to Holland with this ship, leaving behind his princess, his slaves and all his Malagasy friends in tears.[71]

GOUDA[72] (1625)

The 800-ton "spiegelretourschip" *Gouda,* built in Amsterdam, was commissioned by the Amsterdam chapter of the VOC between 1620 and 1625. Under the command of Gerard Douweszoon, her maiden voyage brought her from Texel to Jakarta from December 1620 to August 1621 and returning in 1622. She sailed again from Texel to Jakarta in 1624. On January 27, 1625, captained by Jan Willemszoon Dijk, she left Jakarta accompanied by the *Middleburg* and the *Hollandia*. The ships, filled with pepper, nutmeg and porcelain on this return journey were six weeks at sea when on March 18, 1625 they were caught in a heavy storm east of Madagascar. When it cleared the following day, they learned that the *Middleburg* had lost all her masts and the *Hollandia* her main mast. The *Gouda* was nowhere to be seen but a large quantity of floating pepper indicated that she was lost.

The *Hollandia* was able to reach the Bay of St. Luce and undertook repairs with the assistance of the local population. They lost the commander, Cornelis Reyetsz, and buried him there. After three weeks, they resumed their homeward journey, though two of the sailors had run off with local women. About a year later, both boarded the *Amsterveen* that had taken relief there. The *Hollandia* arrived on November 16, 1625 in Vlissingen. The *Middleburg* had also tried to reach the Bay of St. Luce but was carried northwards by currents and eventually took relief in the Bay of Antongil. They spent about 6 months there, undertaking the needed repairs and loading fresh sup-

Figure 38. The *Gouda, Middelburg* and *Hollandia* caught in a storm off Madagascar on March 18, 1625

plies. She then sailed to St. Helena but was lost shortly thereafter in a fight with the Portuguese forces close to St. Helena.

KONING DAVID[73] (1639)

The 360-ton yacht *Koning David* had a long and varied history. Built on the wharf in Hoorn, a port in North Holland, and mounting 14 guns she was commissioned by the VOC in 1623. Her maiden voyage under the command of Jan Thomaszoon and commission of the trading chamber of Hoorn carried her from Goeree in April 1623 to Batavia (Jakarta) which she reached in late August 1625. She undertook another voyage under the command of Koert Geurtszoon Visser again for the trading chamber of Hoorn from Masulipatnam in March 1626 via the Cape of Good Hope to Maas which she reached in December 1626.

More noteworthy, the *Koning David* sailed from Texel on December 12, 1630 for the South River (the Delaware) to plant the first colony there. Among the early founders of colonies on the banks of the Hudson was David Pieterszen de Vries who was from Hoorn. In the year 1630, "de Vries was associated with De Laet, Van Rensselaer and other patrons to plant colonies within the limits of the New-Netherlands. Their first enterprise was to the South or Delaware River. Towards the close of the year about thirty emigrants started a settlement near the present site of Lewistown, Delaware. The voyage of De Vries cradled a state. Delaware exists as a separate commonwealth due to the colony of De Vries."[74]

After returning to Holland, de Vries went back to the South river in the autumn of 1632. He had the misfortune to find his colony destroyed and laid waste. There was not a solitary survivor to make known their fate. However, it was apparent that it was the work of neighboring Indians. After trying to ascertain the perpetrators of the horrid tragedy, he sailed for Virginia and to the New-Netherlands where he remained until the summer of 1633. The following year he engaged in establishing a colony on the coast of Guyana in South America. This also proved unsuccessful so he abandoned the attempt and went to the Hudson and New Amsterdam in June 1635. He returned to Holland the same year.

In August 1638, the Board of the VOC (Heren XVII) resolved to reinforce the settlement in Batavia by sending around twenty vessels carrying about 2,000 sailors, more than 1,200 relief troops, a considerable load of silver and gold coins and ingots, and supplies. The Amsterdam and Zeeland chapters of the VOC were the main purveyors, assisted by the chambers of Henkhuizen, Delft, and Hoorn. On March 12, 1639, the chamber of Zeeland sent the 200-ton fly-boats *Kapelle* and *Pauw,* and the *Koning David,* the latter captained by Jan Taaikaas. The *Kapelle* left from Goeree, and the two other vessels from Wielingen. The *Kapelle* arrived in Batavia on September 20 of the same year, the *Pauw* on November 12, but the *Koning David* never made it; she wrecked on the Coast of Karimboly in the south of Madagascar on August 29, 1639.

About a month after departing from Holland, the three vessels refreshed in the Cape Verde Islands and thence headed south. The *Koning David* got separated from the two other vessels in heavy seas close to the Cape of Good Hope on June 21. Still sailing alone, on July 16 the *Koning David* was taken in a very violent storm. The fierce winds broke her main mast, which fell overboard. The mizzen-mast got cleaved to the point of not being able to bear any sails. And the fore-mast was so wobbly that they only dared to use the studding-sail boom and the sprit-sail. The storm lasted for a whole week. The crew, drenched by the squalls and the seawater, was shivering and sick with scurvy. At the end of the week, only 8 men managed to operate each watch, all the others were sick in bed.

When the storm blew away, they were so destitute that the decision was taken on August 5 to head for Madagascar. They aimed for the island's southeastern corner and saw land on August 16. They then shaped their course westwards along Madagascar's southern shores, their target being the Dutch Cemetery (Bay of Ampalaza and Nosy Manitsa). But, on August 28, they saw a large bay close to the mouth of a river. They swung out the cutter, the pilot explored the bay, and reported the seafloor to be good, 8 or 9 fathoms deep, and favorable for an anchorage with the prevailing south-south eastern winds.

The next morning, August 29, 1639, the battered *Koning David* entered the bay with a very light southeastern wind. The crew found out that the seafloor was not firm at all, but muddy and slick. Shortly after anchoring, the winds suddenly switched directions, now blowing from the west to southwest, with rising force, and the swells became very heavy. By two in the afternoon, despite having furled all the sails and placed the yard on deck, the cable of the main anchor at the bow ripped, and the chain of the small anchor could not hold. The *Koning David* was pushed on top of four pieces of rock and a few sharp reefs, which shredded the last cable. She drifted between two sandbanks, hit rocks lying close to shore, burst open and foundered.

The violence of the crash had been such, that the main body of the vessel was lying a few feet to the fore of the keel, which was firmly wedged in the rocks. In no time the *Koning David* would be broken up by the brutal surf. The about 400 crew and passengers scrambled for shore. All managed to reach the coast, except for 11 men, including the two surgeons who drowned in the waves. They were now stuck close to the mouth of a river (Manabovo, a little to the east of False Cape). Not an ideal place, by far: the river's output flow was insufficient, the forest was far away, and the sun scorching. The only food they could save from the ship were a few barrels of salted meat and bacon, some vinegar and two olive jars.

An encampment was set up at the mouth of the river using sail, oars and debris from the *Koning David*. Some survivors went to fell trees in the vicinity and, instructed by the ship's carpenters, sawed the wood needed to build a 40 foot long boat. Others built a small fort that was half earth and half wood. Others walking inland bought cattle and other supplies from the natives with coins they had salvaged. They brought the cattle to the fort and placed them in pastures nearby. But, after nightfall the natives raided the cattle they had sold that afternoon. Lured by the coins they started harassing the survivors. A number of skirmishes and retaliatory raids into the villages ensued. In one of those, Captain Taaikaas and a few others were killed.

After several months of work, they finished building the long boat, in which 33 passengers could fit. All of the officers and some of the men boarded her on January 5, 1640 and headed for the Cape of Good Hope, steered by the pilot, taking with them a few muskets and some provisions. After encountering many momentous dangers, they managed to arrive safely on February 9. The were taken on board of the *Maria de Medicis,* the *Breda* and the *Wesel* and brought back to Europe.

Before departing, they had promised to return for those staying behind as soon as possible; a few months at most. They had also given them the rest of the salvaged silver coins; about 200 to 300 pieces of eight for each man. Time went by at the mouth of the Manambovo river fighting hunger, the natives, and the mosquitoes. Many of the near 300 survivors who remained behind died from fevers or from Malagasy spears.

The thinning band of survivors, tired of lingering beyond the waiting period, decided to leave this wretched place and head for the Bay of St. Luce. They dispersed in small groups and went inland. Some died from tropical diseases or from hunger but most were ambushed and killed by the Antampatrana tribes (inhabitants of the Androy plain) to pilfer their silver coins. None of the Dutchmen saw their home country again. The only one who came home was a French sailor. Two French crew members made it to the Anosy region (around Fort Dauphin) where they were employed by King Andriamasikara as foremen. On the way they buried around 600 pieces of eight but were too scared to return and fetch them.

After one died, the King allowed the other to board a French ship returning from the Red Sea. In November 1642, Captain Gilles de Régimont had sailed to Madagascar, arriving on May 1, 1643 at Sainte Luce. There he met the stranded sailors of the *Saint Alexis,* including François Cauche.[75] On August 15, 1643, Régimont weighed anchor for a buccaneer run in the Red Sea and Cauche embarked with him. Finding luck, Régimont captured a rich prize, and put into Sainte-Luce on the way back arriving in early November 1643. His ship left Madagascar in March 1644, with the sole survivor of the *Koning David.* She arrived in Dieppe on July 21, 1644 after a favorable crossing.

TULP (1655)

The galliot *Tulp,* built in Amsterdam, left Texel on December 23, 1653 for her maiden voyage under the command of Captain Samuel Volkertsz. She was commissioned by the Amsterdam chamber and destined for Cape Town. On board were oranges, lemon and pummelo that were introduced to Southern Africa for the first time. Upon her arrival on April 18, 1654 this was recorded in the daily journal of events kept by van Riebeeck, the first governor of the Dutch colony at Cape Town.[76] The *Tulp* brought the first sweet orange trees from the island of St. Helena which were planted in the governor's private garden. In July 1661, the first fruits ripened and were plucked and tested by the governor and found "to be good." The *Tulp* remained in the service of Governor van Riebeeck. She was the only supply vessel that he could use as the one assigned to that task was unable to berth at Tafelbaai.

In June 1654, food supplies fell to troublingly low levels at the colony. The Governor decided to send the *Tulp* to fetch rice in Madagascar. The command of this expedition was given to Frederik Verburgh, the second in command at the Cape colony.[77] The *Tulp* cast off in early July 1654 and spent a month in the Bay of Antongil where they could procure rice. The crew was hosted by the local king after the ship's doctor saved him from poisoning. Starvation of the Cape colony was only averted by her timely return on December 12.[78]

Governor van Riebeeck sent Frederik Verburgh on a second supply mission to Madagascar to secure rice, beans, wax, ivory, hides and slaves. He asked him to make as many geographical and nautical observations as he could as he wished to engage in trade with Madagascar once he received a bigger ship from Holland. But the *Tulp* was to sail to Mauritius first to deliver a stern letter to the Governor of Mauritius Maximilien de Jong. In the letter van Riebeeck admonished him to improve the management of the island and to abandon his idea of renewing trade with Madagascar.[79] He also asked him to hand over the journal and the papers of former Governor van der Stel who had passed away.[80]

The *Tulp* left Table Bay on August 14, 1655. Nothing was heard of her for 18 months even though the galliot *Nachtglas* was sent from the Cape in July 1656 to look for her. When the French ship *Le Maréchal* took relief at Table Bay on March 31, 1657, bringing back four seamen of the *Tulp,* did her wreckage become known. The seamen's testimony was reported in van Riebeeck's journal.[81] After putting in at Mauritius, the *Tulp* went to Antongil to load rice. Then she sailed to the Bay of Loky (in the Northeast) to load more rice. This is where Verburgh got word that the King of Fenerive (Fenoarivo Atsinanana) asked to see him. Eager to establish good relations with this King and hopefully weaken the growing French presence in Madagascar, Verburgh sailed from Loky on November 23, 1655 and arrived in Fenerive a few days later.

During the night of December 2, 1655 the *Tulp* was caught in a violent cyclone, dragged her anchors and was carried into the open sea. With the wind blowing harder and harder, the crew cut the main mast and she floated helplessly until being thrown on the coast at Fenerive where the ship and cargo were lost. The next day, the crew salvaged some of the cargo and remained close to the wreck for 8 days. Some of the local population helped them to transport the salvaged goods to a nearby village where they stayed for about a month. The King of the Island of St. Marie heard that a Dutch ship had wrecked and came with four canoes. He brought the pilot and some of the crew to St. Marie. Verburgh and the rest of the survivors joined them a few days later. The King of St. Marie treated them well. Verburgh bought four houses in one of the villages where they settled. Unfortunately, an epidemic broke out and 13 of the survivors died in a few days, including Frederik Verburg, the pilot Cornelis Janssen Holsteyn and the bookkeeper Cornelis van Heyningen.

In May 1656, four French vessels came to St. Marie, the *Duchesse, l'Armand, Le Maréchal,* and the *Saint-Georges*.[82] The *Maréchal* went cruising in the Red Sea and returned to St. Marie in December 1656 where she found two of the other ships in a very bad condition. The *Armand* and the *Saint-Georges* were declared unseaworthy and beached. In January 1657, the 10 remaining survivors of the *Tulp* wreckage were taken aboard the *Duchesse,* destination Fort-Dauphin, which they reached two weeks later. The *Maréchal* joined them soon thereafter, having stayed behind to pick up the artillery and some debris of the two abandoned ships. The *Duchesse* was also declared unseaworthy and beached in Fort Dauphin. The *Maréchal* left Fort Dauphin in late February 1657 but the Captain could only take four of the survivors of the *Tulp*. He already had so many people on board that he had to leave some of his own crew behind. The remaining six survivors stayed behind in Fort Dauphin to guard the salvaged goods.

Four years later, in February 1661, Jacques de Bollan sailed from Batavia with the flute *Postillon,* which was rented from the VOC to pick up Mr. Du Rivaux, the French Governor of Fort Dauphin and the cannons from a wrecked ship.[83] The *Postillon* left Fort Dauphin in May, planning to put in at Mauritius but adverse winds pushed her back toward Madagascar. She first touched land at the Island of Saint Marie thence put in at the Bay of Antongil on June 17, 1661. There, Bollan encountered three survivors of the *Tulp* who had not gone to Saint Marie with the others. He took them on board and brought them back to Batavia.[84]

ZEELT[85] (1672)

The small 90-ton hooker *Zeelt* was built in 1668 in the Rotterdam shipyard for the Rotterdam chamber and was used by the VOC from 1669 to 1672. She left for her maiden voyage under the command of Theunis Gijsbertsz on January 11, 1669 from Maas. She put in at the Cape settlement in May and reached Batavia on August 9, 1669. She went missing after leaving the Cape in 1672.

A few months later, the crew of the *Grundel,* which left the Cape settlement on October 4, 1672 destined for Mauritius was taking on fresh water close to Fort Dauphin in southern Madagascar. Suddenly, they were attacked by 80 Frenchmen and about 100 local warriors. The commissioner and a sailor were taken prisoners. The Captain and 11 men, three or four of which were hit by artillery, escaped in their canoe. At the same spot, the carcass of a ship lying on the beach looked like the *Zeelt* that had gone missing. Some of the men aboard the *Grundel* had sailed with the *Zeelt* before and positively identified her. Three weeks after leaving Madagascar the *Grundel* wrecked a few miles from False Bay (Simonstown in South Africa).

RIDDERSCHAP VAN HOLLAND[86] (1694)

The *Ridderschap van Holland,* built in 1681 in Amsterdam and commissioned by the chamber of Zeeland, belonged to the largest class of the company's ships. 160 Amsterdam feet (45.3 meters) long with a nominal carrying capacity of 520 tons and an actual capacity of 1,138 tons. She undertook four successful voyages between Holland and Batavia between 1683 and 1692. She left Wielingen for her fifth and final voyage to the Indies under the command of Captain Dirk de Lange on July 11, 1693 with 325 passengers on board including number of VIPs, of which a senior VOC official, Sir Lames Cooper (originally from Scotland), who was scheduled to take up an appointment as a member of the Council of the Indies in Batavia. She was also carrying considerable funds on account chamber of Zeeland, reported as more than one ton of silver and gold coins and ingots, not to speak of the rich personal belongings of the VIPs on board.

After an uneventful voyage down the Atlantic Ocean, the *Ridderschap van Holland* called at the Cape settlement on January 9, 1694. The ship disappeared after leaving the Cape colony bound for Batavia, on February 5, 1694. Early conjecture was that the *Ridderschap van Holland,* like the *Batavia* and the *Vergulde Draak* before, had probably wrecked on the coast of the South Land. In 1696, it was decided to dispatch a search expedition of three vessels, under the command of Willem de Vlamingh, to seek signs of the wreck and rescue any people who might have survived.

Rumors were later received at the Cape to the effect that the *Ridderschap van Holland* was taken by pirates based at Fort Dauphin near the southeastern corner of Madagascar. In June 1698 the Cape

Political Council recorded that the ship might have been lost in that way. Even so there was hope that survivors could be found. In early May 1699, the yacht *Tamboer* left Tafelbaai. On its way to Batavia she called on Fort Dauphin in early July to establish the fate of the missing vessel. Captain Jan Coin did extensive interviews with elder local inhabitants and some European settlers.[87] He learned that about four years earlier (1694) people had found large wooden beams, masts and fourteen graves at Manantena about 7 or 8 leagues north of Fort-Dauphin. Also, around the same time, large pieces of wood and fragments of masts had washed ashore close to Star Bank (Banc de l'Etoile), but no survivors were found. There was no proof however, that the *Ridderschap van Holland* had wrecked in either one of these places. Based on further research, this author believes the Banc de l'Etoile to be the probable place for this wreckage.

REFERENCES

[59] Guinote, op. cit, p. 16; Meilink-Roelofsz, M. A. P., "The structures of trade in Asia in the sixteenth and seventeenth centuries. Niels Steensgaard's «Carracks, caravans and companies». The Asia trade revolutions. A critical appraisal" in *Mare Luso-Indicum,* 1980; s.n., *L'Océan Indien, les pays riverains et les relations internationales. XVIe-XVIIIe siècles;* Paris; Société d'Histoire de l'Orient; vol. IV; pp 1-40, 1962, p. 135 ; Sabrizain, "Sejarah Melayu, A history of the Malay Peninsula," at http://www.sabrizain.demon.co.uk/malaya/dutch.htm.

[60] Major, R.H., "Early Voyages to Terra Australis, now called Australia," a Gutenberg of Australia eBook No. 0600361.txt produced by: Col Choat and Bob Forsyth, http://gutenberg.net.au, April 2006.

[61] See, *inter alia,* Sabrizain, "VOC: The first multinational" in Sejarah Melayu, A history of the Malay Peninsula, at http://www.sabrizain.demon.co.uk/malaya/dutch.htm.

[62] Though the Portuguese discovered Mauritius in 1505, they generally neglected it, allowing Holland to move in on the island. The second expedition of the Old Company landed there in 1598 under the command of Admiral Wybrand van Warwyck and claimed the place for Maurice of Nassau, then Holland's head of state, naming it Mauritius after him. However, only in 1637 did the VOC board decide that a permanent settlement was needed to prevent the French or the British taking it over. Moreover, it would be a useful refreshing station on the way to Indonesia as food and timber was plentiful on the island. In 1638, a group of colonists was sent and a fort built. The first slaves used by these settlers came from Madagascar. However, the Dutch colonists failed to

find a foothold on the island, even though they drove a large native bird, the dodo, into extinction. The first settlement lasted twenty years but was abandoned in 1658. The Dutch then sent new colonists and this cycle went on and on until the Dutch abandoned their possession of Mauritius in 1710.

[63] Several works relate the early expeditions of the Old Company; for a review, see Voorcompagnieën, The VOC site, http://www.vocsite.nl/ and Grandidier et al., op. cit., T. I, pp. 163-165;

[64] "Het tweede Boeck, Joernael uit Dagh-register inhoudende een warachtig verhael ende historische vertellinghe van de reyse, maart 1598," ed. Bernard Langenes, Middelbugh, 1601; Grandidier et al., op. cit., t. I, pp. 240-248.

[65] De Constantin, "Recueil des voyages qui ont servi à l'établissement et aux progrès de Compagnie des Indes Orientales formée dans les Provinces Unies des Pays-Bas," 1725, t. III, pp. 352-362; Grandidier et al., op. cit., t. I, pp. 257-263.

[66] Source: The VOC site, http://www.vocsite.nl/; De Constantin, op. cit., t. IV, p. 93.

[67] These must have been descendents from part of the crew of the *São Ildefonso* that shipwrecked in 1527 and settled in St. Luce (see above), possibly joined by some of the crew of the *Conceição* and *São Sebastião*. The story is related by João de Barros, op. cit., Dec. IV, 1613, liv. III, ch. II, pp. 263-264, Grandidier et al., op. cit., t. I, pp. 265-268, and Canitrot, "Les Portugais sur la côte orientale de Madagascar et en Anosy au XVIe siècle (1500-1613-1617)," Revue de l'Histoire des Colonies, pp. 208-209.

[68] Letter by the Augustine friar Athanase to Archbishop Dom Frei Aleixo de Meneses, reproduced in Boletin da Sociedade de Geographia de Lisboa, 1887, pp. 334-335.

[69] Sources: The VOC site, http://www.vocsite.nl/; Mulder, W.Z., "Hollanders in Hirado, 1597-1641," Haarlem: Fibula-Van Dishoeck, 1985; *RGP-GS166,* "Dutch-Asiatic Shipping in the 17th and 18th centuries, Volume II, Outward-bound voyages from the Netherlands to Asia and the Cape (1595-1794)," Den Haag: Martinus Nijhoff, 1979; *RGP-GS167,* "Dutch-Asiatic Shipping in the 17th and 18th centuries, Volume III, Homeward-bound voyages from Asia and the Cape to the Netherlands (1597-1795)," Den Haag: Martinus Nijhoff, 1979; Valentijn, F., "Oud en Nieuw oost-Indiën, deel IV/A," Franeker: Van Wijnen, 2003.

[70] According to Valentijn, op.cit., the skipper was Jakob Franszoon.

[71] Flacourt, "Histoire de Madagascar," 1638, pp. 37-39; Grandidier et al., op. cit., t. II, pp. 287-290. Flacourt mentions that Pieter's wreckage took place 'around 1618,' but no Dutch ship is recorded to have wrecked off Madagascar around that date. It is not clear which ship brought Pieter back to Holland. Given the reported seven years since Pieter's wreckage, it must have been around 1620. According to Flacourt, it is a ship on the homebound voyage. The only recorded berthing in St. Luce around that time was an outbound voyage (the *Zierikzee* in 1617). Also, Willem IJsbrandtsz Bontekoe aboard the *Hoorn* at-

tempted to anchor there on its ill-fated outbound voyage in 1619 (the ship blew up west of Sumatra), but could not anchor because of heavy seas and then took relief on the Island of Mascareigne (Reunion).

[72] Sources: The VOC site, http://www.vocsite.nl/; Bontekoe, W.Y, "Het Journael of de gedekwaerdige beschrijvinghe and de Oost-Indische Reyse van W.Y. Bontekoe van Hoorn, begrijpende veel wonderlijke en gevaerlijcke saecken hem daar in wedervaren," Amsterdam, 1630; Miller, R., "De Oostindievaarders," Amsterdam: Time-Life Boeken, 1981; Voyage Nr. 5168.2, Gouda, at http://www.vocshipwrecks.nl/home_voyages/gouda.html; Bruijn, J.R., Gaastra, F.S., Schöffer, I., "Dutch-Asiatic Shipping in the 17th and 18th Centuries," The Hague, 1979, 1987; Bostoen, K., Daalder, R., Roeper, V., Verhoeven, G., and Wildeman, D., "Bontekoe. De schipper, het journaal, de scheepsjongens," Zutphen, 1996.

[73] Sources: The VOC site, http://www.vocsite.nl/; various correspondences to the Board (Heren XVII) of the VOC in Amsterdam, including a letter by the Governor General of Batavia dated January 31, 1640, and a letter by Nicolaes Koeckebaeker, Captain of the *Nieuw Amsterdam,* dated July 20, 1640; Flacourt, E., "Histoire de la Grand Isle Madagascar," Paris, 1658, pp. 33, 36 and 37 (Flacourt erroneously mentions that the wreckage took place « around 1635 »).

[74] Bancroft, G., "History of the United States of America, from the Discovery of the Continent," New York: D. Appleton and Co., 1896, ii. 381.

[75] See below under the *Saint Alexis.*

[76] In 1652 about ninety men led by Jan van Riebeeck had established a refreshment station for the VOC at the Cape between the foot of Table Mountain and the shores of Table Bay. The purpose was to provide fresh water, fruit, vegetables, and meat for passing ships en route to India, as well as to build a hospital for ill sailors. Vegetables and fruit were cultivated to help prevent sailors from getting scurvy. See Webber, Herbert John, "A comparative study of the citrus industry of South Africa," So. Africa Dept. Agr. Bul. 6, 1925, p. 10.

[77] Verburgh's report was recorded in the Governor's journal; see Leibbrandt, H.C.V., "Précis of the Archives of the Cape of Good Hope: Riebeeck's journal," Cape Town: W. A. Richards & Sons, 1897, vol. I, pp. 206-207.

[78] Worden, N., Van Heyningen, E., Bickford-Smith, V., "Cape Town: The Making of a City: An Illustrated Social History," Cape Town: Verloren, 1998, p. 19.

[79] After Governor van der Meersch's third voyage to Madagascar in 1647 and the folding of the settlement in the Bay of Antongil, the governors of Mauritius were not anymore authorized to trade with Madagascar.

[80] Letter from van Riebeeck to Maximilien de Jongh, August 10, 1655, Rijksarchief, Den Haag.

[81] Leibbrandt, H.C.V., op. cit., vol. II, pp. 55-57.

[82] For a description of this fleet, see below under the French East India Company.

[83] Most probably the *Duchesse.*

[84] Dagh-Register gehouden int Casteel Batavia vant passerende daer ter plaetse als over geheel Nederlandts-India, Anno 1624-1629, 1661, p. 306, and 's-Gravenhage: Nijhoff, 1887.

[85] Sources: van Dam, Pieter, "Beschrijvinge van de Oostindische Compagnie, eerste boek, deel I." 's-Gravenhage: Martinus Nijhoff, 1927; Dagh-Register gehouden int Casteel Batavia, 1673, p. 171.

[86] Sources: VOC site http://www.vocsite.nl/ and http://www.vocshipwrecks.nl/; Bruijn, J.R., Gaastra, F.S., Schöffer, I., "Dutch-Asiatic Shipping in the 17th and 18th Centuries," The Hague, 1979, 1987.

[87] His report can be found at the Rijksarchief, Den Haag, Koloniaal Archief 4020, Tweede deel der brieven en papieren van de Kaap [de Goede Hoop] overgekomen, ff. 820-825.

3

The British East India Company[88]

The Honorable British East India Company (HEIC), sometimes referred to as "John Company," was one of the first joint-stock companies preceded only by the Dutch East India Company. It was granted a Royal Charter by Elizabeth I on December 31, 1600, with the intention of favoring trade privileges in India. This was renewed in 1609 by King James I for an indefinite period. The Company transformed from a commercial trading venture to one that presided over the creation of the British Raj and virtually ruled India as it acquired auxiliary governmental and military functions until its dissolution in 1858.

There were two attempts made by the English to open trade with the east before the incorporation of the East India Company. The first expedition consisted of three ships that sailed from Plymouth for the East Indies on April 10, 1591: the Admiral ship *Penelope,* with Captain George Raymond; the *Royal Merchant,* with Captain Abraham Kendall; and the *Edward Bonaventura,* with Captain James Lancaster. On their arrival at the Cape of Good Hope the *Royal Merchant* was obliged to return to England. After passing Cape Corrientes, at the entrance of the Mozambique Channel, the other two ships were separated in a violent storm and the Admiral ship *Penelope* was never heard of again. After passing by the northwest of Madagascar, while waiting for the *Penelope* in the Comoros the *Edward Bonaventura* was hit by lightning which caused severe damage and loss of several lives. Moreover some of the crew went ashore and were murdered by the local population.[89] Captain Lancaster proceeded to India in September 1591. He returned from Ceylon in December 1592 following the Maldives and then the east coast of Madagascar and reached St. Helena on April 3, 1593. From there, he was compelled by distress to steer to St. Domingo where his ship drove out to sea with only five men on board. Captain Lancaster arrived in England on May 24, 1594, in a French vessel.[90]

The second voyage was equally unfortunate. Out of three ships, which sailed from England in 1596, not one returned. No further attempt was made by the English to open trade with the east until the incorporation of the East India Company.

The HEIC was founded as *The Company of Merchants of London Trading into the East Indies* by a coterie of businessmen who obtained the Crown's charter for exclusive permission to trade in the East Indies for a period of fifteen years. The Company had 125 shareholders and a capital of £72,000 (equivalent to $180 million today). Part of the initial capital was immediately laid out in the equipment of five ships, the *Dragon,* the *Hector* (Captain John Middleton), the *Ascension,* the *Susan* and the *Guest.* All were under the command of the same Captain James Lancaster who led the 1591 expedition. The fleet passed the Cape of Good Hope on November 1, 1601 and the northern point of Madagascar on November 26 sailing to Mauritius. Contrary eastern winds prevented them from reaching Mauritius. They put in first at the Island of St. Marie and then at the Bay of Antongil. They wintered there and left for India on March 6, 1602. This opened the Indian trade for the British. Captain Lancaster returned to England on September 11, 1603. The third HEIC fleet also put in at Madagascar: the *Consent* commanded by David Middleton anchored in the Bay of St. Augustine on August 30, 1607. The two other ships, the *Hector* and the *Dragon,* commanded by Captains William Keeling and Hawkins, respectively, put in there on February 17, 1608.

Initially, the HEIC had little impact on the Dutch control of the spice trade. At first it could not establish a lasting outpost in the East Indies. Progressively, however, trade transit points were established in Surat and in the town of Machilipatnam on the Coromandel Coast in the Bay of Bengal. England finally gained foothold on mainland India in 1615 by signing a treaty with the Mughal emperor Jahangir who ruled most of the subcontinent. This gave the Company exclusive rights to reside and build factories in Surat and other areas. In return they offered to provide the emperor goods and rarities from the European market. This let the HEIC eclipse the Portuguese Estado da India and make inroads into the Dutch monopoly of the spice trade. By 1647 the Company had 23 factories each under the command of a 'factor' or master merchant and 90 employees in India.

In 1657 Oliver Cromwell renewed the charter of 1609 and made minor changes in the holding of the Company. Its status was further enhanced by the restoration of monarchy in England. By a series of five acts around 1670, King Charles II provisioned it with vast rights so that the HEIC arguably became a "nation" in the Indian mainland. It independently administered presidencies and possessed a formidable military strength. However, under pressure from ambitious tradesmen and former associates of the HEIC (pejoratively termed *Interlopers*) who wanted to establish private trading firms in India, a deregulating act was passed in 1694. This allowed any English firm to trade with India unless specifically prohibited by act of parliament. This annulled the charter that was in force for almost 100 years. By an act passed in 1698, a new "parallel" East India Company (officially titled the *English Company Trading to the East Indies*) was floated. However, powerful stockholders of the old company quickly subscribed in the new concern and eventually dominated the new body. After wrestling with each other for some time the companies merged in 1702. The amalgamated company became the *United Company of Merchants of England Trading to the East Indies*.

Many HEIC fleets visited Madagascar on their way to India with the Bay of St. Augustine being the spot of choice to secure refreshments and care for the sick. A small English settlement was established in the Bay of St. Augustine in 1644 and another one on the Island of Nosy Be but both were quickly abandoned. Nonetheless, most HEIC fleets preferred to take the outer passage to India, i.e. rounding Madagascar to the east and refreshing in Mauritius which was a regular port of call for HEIC ships going to and from India. As a result only four HEIC ships are recorded to have wrecked off Madagascar.[91]

ANNE[92] (1689)

The 120-ton *Anne,* mounting 18 guns and manned with a crew of 25 was licensed under East India Company regulations in 1688. She

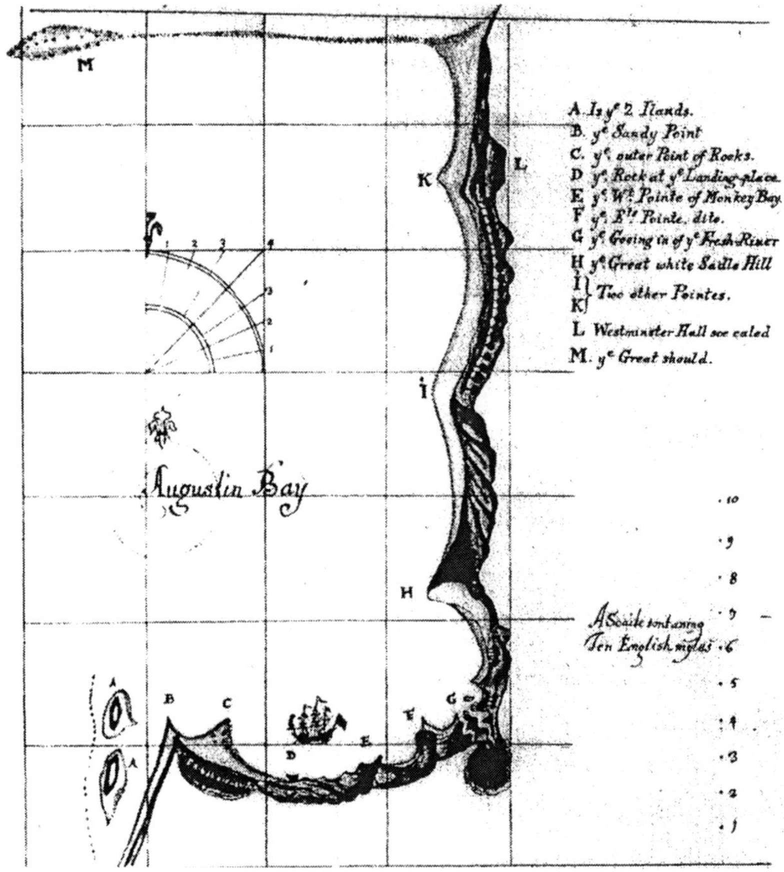

Figure 39. Plan of St. Agustine Bay, 1650 (Source: Manuscript by Charles Wilde, "Journal kept by mee Charles Wilde, purser in Shipp 'Bonitto' being bound by God's Assistance from England to the island Assada on the West Side of St. Laurence and from thence to Fort St. George at Madraspatam on the coast of Choromandell, Michaell Yate Comander, 2 June 1650-28 July 1652").

Figure 40. Lateral view of St. Augustine Bay, 1650 (Source: Charles Wilde, op. cit.)

left London in February 1688 on her maiden voyage for the Company, under Captain William Freke accompanied by the *Defence*, the *James*, and the *Mary*. She arrived in Morondava in late 1688. The

Figure 41. Saint Augustine Bay today

Figure 42. Saint Augustine Bay today

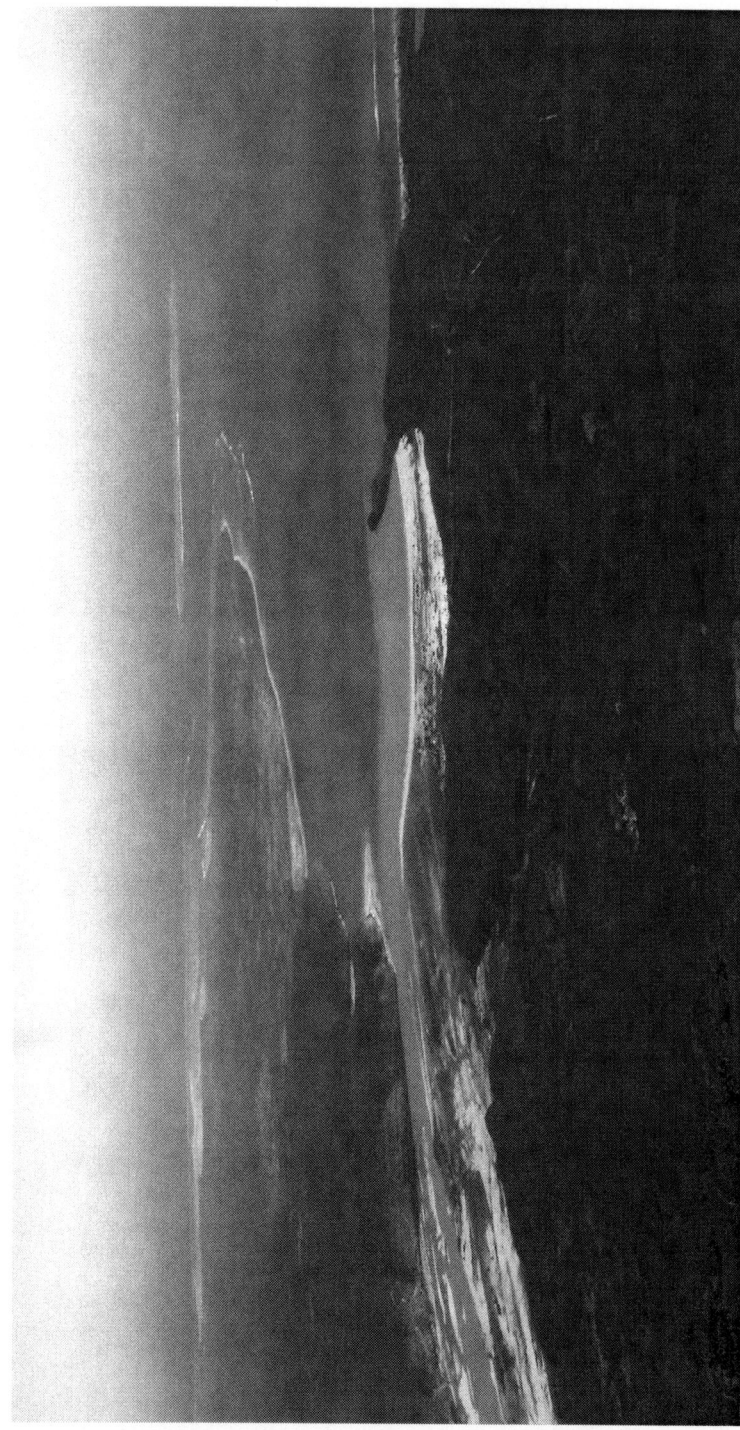

Figure 43. Saint Augustine Bay today

three other ships proceeded to India. Captain Freke visited the King of Morondava and contracted for a cargo of slaves to be ready three months later. The *Anne* then sailed in early 1689 to the Delagoa River (now called Maputo Bay on the southeast coast of Mozambique) to undertake some other trade. On April 24, 1689 she sailed back for Madagascar to take in the cargo of slaves that was supposed to be ready.

Captain Freke and the first mate were bedridden with fevers and unable to perform. The charge of navigating the ship lay on the second mate. Due either to the second mate's ignorance or adverse conditions it took twice the usual time to cross the Mozambique Channel. St. Laurence was sighted on May 9 in the evening but the ship was still 50 leagues south of Morondava. The next day the *Anne* sailed on a northward course along shore and had good soundings. Towards evening though, they were almost out of sight of land. The second mate was ordered to steer northeast towards the coast till he had depth of 15 fathoms but not less and at 8 at night he had that depth. However, the second mate kept steering in towards the shore. By 10 pm he had shouldered the coast so much that he was ashamed to discover the low depth he had steered into. Though the night was without cloud and he was using only the foresail for fear of overshooting the port, either by carelessness or ignorance, he ran the *Anne* ashore on an island 3 miles from Morondava.

After hitting the islet all hands spent two hours trying to heave the ship off. However, she took on too much water. When the cabin was half submerged, they despaired saving her and made haste to save their lives. They all made it saving only the clothes on their backs except for a young African boy who was sleeping in the hold and drowned. The next morning, May 11, the weather was fair at low water. The men saved their clothes that were not washed out of the ship the night before at high water and whatever money they could find. They also saved 6 gallons of water and lots of strong liquor which they soon abused.

A row erupted between Captain Freke and the inebriated men. The Captain implored the men to go to the mainland to get water and provisions, in vain. The men refused to leave the small islet without first receiving a share of the money that was saved. They argued that they had as much right to it as the Captain. If denied, they threatened to maroon the Captain and the men who stood with him. Those were the chief mate, purser, surgeon and cooper all of whom were sick. After a good nap the row quieted down. The next day, May 12, all agreed that it was in vain to stay in a place where there was no water nor provisions. They all embarked on the small sloop and the longboat and sailed to the port of Morondava.

They were civilly received and supplied with provisions. After considering their options, the officers decided that the best course of action to return to England was to sail to the Island of Anjouan (Comoros) and wait for some East Indiaman to touch there. Having stocked the sloops with plenty of provisions they sailed for Anjouan on May 23 arriving there on May 31. In the morning of June 9 they spied two sails making for Anjouan. Their hopes were dashed however when they found out that the ships were not East Indiamen but two pirate ships; the *Cygnet* (renamed *Little England*), Captain Teat; and a brigantine, Captain Knight (see story below). Though Captain Freke used all the arguments he could to dissuade them, eight crew members of the *Anne* shipped themselves on board the pirate ships which sailed northwards for the Red Sea on June 11.

The remaining crew, many of whom fell ill, waited in vain until July 20, 1689 for a ship to arrive. Despairing to see a ship until the next year they decided to fit the small sloop for the passage to India. After provisioning her and putting aboard their possessions they sailed on August 2 leaving one man behind in Anjouan. The sloop proved leaky so they touched at the small island of Pate (off Praslin, Seychelles) to caulk her before proceeding to Oman. There they were well treated by the Governor of Muscat. They learned the unfortunate news; that the Mogul army was occupying the Island of Bombay. This put them in great doubt as to what course to take. They finally decided to set their course for the Vergula shore (on Maharasthtra's Konkan coast). They anchored close to some rocks on September 14 but found out they were still considerably short of Vergula. They continued southward the next morning but adverse winds forced them to anchor again.

Not sooner had they come to an anchor that two small boats came out to them from behind a point of land that lay southward. These were Malabar pirates who lay there to take small vessels that passed that way. With contrary winds preventing them to put out to sea, the English decided to go on shore and fight the pirates with the three pounds of powder and 100 cartridges they had on board. The pirate sloops anchored a distance away from the surf and out of musket shot range. They fired a volley of small shot that did no damage. The English answered it and had the good fortune to kill a pirate. The pirates continued to fire their small arms damaging the sloop's rigging. With the tide ebbing the pirates launched an assault.

The English crew strongly resisted at first killing two more pirates including their Captain. However about 300 pirates pressed on. Around one in the afternoon, the English were in a desperate situation having only 20 powder cartridges left. The pirates sent an emissary who told them that if they were willing to surrender the sloop, goods and money they would give them a pass and that they would have the liberty to go wherever they pleased. The English refused to surrender the boat. The emissary left and soon returned with a pass with two seals on it which he told them was sufficient to secure them from abuse in their travels. Surrounded by the pirates and nearly out of ammunition Captain Freke reluctantly surrendered the sloop. The pirates immediately took possession of her and plundered the English of everything save the clothes on their backs.

At daylight on September 16, the pirates told the English that they wanted to ransom them and forced them to march for two days to a fort. After some days rest they were marched to another fort. Realizing that there would be no redemption as long as the war continued between the Mogul and the British, Freke plotted an escape to the Dutch factory at Vingula. That was a few days' march away and most did not wish to attempt it. Leaving eight crew members behind, Captain Freke and the surgeon escaped on October 5 and marched to Vingula which they reached following much hardship and hunger. From there a canoe took them to Goa and a few days later they got a passage to the English fort at Karwar. They remained there until November 20. Then they took a passage on the *Hossanna* and proceeded to Bombay Castle before returning to England.

DEGRAVE (1703)

The *Degrave,* a 520-ton, 52-gun English East Indiaman built in 1698 set out under the command of Captain William Young from the Thames outside London on February 19, 1701 bound for Bengal. This was her second voyage to India. She had gone in March 1699 and returned in 1700. This well-armed merchant ship carried a crew of at least 112, twenty of whom were on the previous voyage, and at least seven passengers. She arrived in Calcutta in June 1701. A combination of nine passengers and crew drowned in Bengal and another 40 died from disease during the stay there including Captain William Young. Command of the ship was given to his 26-year old son Nicholas. The *Degrave* was due to leave in early 1702 but was ordered to remain in Hooghly to protect the East India Company factory during a period of conflict with the native government. She set sail from Bengal on Christmas Day 1702 for the return voyage to London with a cargo of textiles.

While departing with her new complement of around 120, including two women and other passengers, the *Degrave* ran aground on a sandbank while moving down the Ganges. She was dislodged by the high tide with no apparent harm to the ship. However, at sea she began to leak badly. Two chain-pumps were used continually over the next two months on her way to Mauritius. She put in there in March 1703 to undertake repairs. There 12 Moors and 30 Banians[93] were taken on board. They were the crew of a rich Arab ship marooned on Mauritius by the pirate Captain John Bowen. The carpenter was unable to locate the breach so she pursued her voyage towards the Cape with the water level in the hold continuing to rise despite the extra labor. Fearing for their lives, the crew compelled the young Captain to make for Madagascar whilst several cannons and heavy goods were thrown overboard to lighten the ship.

They got sight of Fort Dauphin but, being aware of the terrible fate that had previously befallen the French colony, they decided to forego the hospitality of King Samuel who was reigning over the Anosy region.[94] The *Degrave* continued to steer westwards along Madagascar's south coast in search of a better haven. However, the ship was so leaky that by then they had 4 feet of water in the powder magazine. As they could not get the vessel dry she was ready to sink. On April 26, 1703 Captain Young decided to anchor in the Tanroy region south of the village of Belitsaky a quarter mile away from a two leagues (6 nautical miles) wide beach and away from breakers. All masts and rigging were cut off and the cannons and merchandise thrown overboard to keep the ship afloat. As the long-boat and pinnace were left behind in Bengal and only a small boat remained, the crew started building a raft with planks and yard to get everyone ashore. Some of the local fishermen, becoming aware of the distress lighted fires on the beach to guide them.

The raft was finished during the night. In the morning of April 27, 1703, Mr. Pratt, the chief mate and four sailors were sent in the boat with a long rope for a warp to fasten on the land. With heavy breakers constantly running upon the rocks there the boat broke into pieces before reaching land. However, being near it and with the help of local fishermen they saved the part of the boat to which the rope was fastened. About 50 of the passengers, including one of the two women on board, climbed on the raft and were hauled by the rope toward the shore. When they reached the breakers, the first wave turned the raft topsy turvy and the passengers were washed off. All reached the shore with great difficulty but the woman who drowned. There was such a surf run that they dared not venture with the raft again. Perceiving this the Captain ordered the cable to be cut and let the ship drive near to the land where she soon was beaten to pieces. He got on shore with his father's heart in his hand. According to his dying request, it was put into a bottle in order to be brought to England and buried at Dover. Eventually, they all got to shore on pieces of the ship except two men who drowned. All in all there were 160 survivors including the Moors and Banians.

In 1729, Robert Drury, a 15-year old midshipman at the time of the wreckage published an account of his 15 year captivity in Madagascar.[95] For years it was dismissed as fiction and attributed to Daniel Defoe as it came out ten years after he published Robinson Crusoe in 1719. However, research by Professor Secord of the University of Illinois[96] and archaeological work undertaken by Mike Parker Pearson, Reader in Archaeology at the University of Sheffield, has proven that it was true.[97]

Survivors of the wreckage were soon surrounded by 200—300 local Tandroy (or Antandroy) people who had gathered on the beach to pick up the ship's cargo that was driven ashore in whole bales. The natives had a particular interest in iron and pieces of silk and fine calico while neglecting the muslin. The survivors remained thus two days and nights without resolving what to do. The next evening, Sam an Englishman, arrived in advance of the king's party. He was one of the eight English and Scots living with the Tandroy. Their party included Robert Drummond, Captain of the *Speedy Return,* and Alexander Stewart, Captain of the *Content,* who were cast away in eastern Madagascar when their ships were captured by the pirate John Bowen.[98] Sam told them that they would be well looked after by the king but might never be allowed to leave.

The crew of 160 men was marched from the coast to the capital, Fenoarivo. Some managed to escape and walked overland to St. Augustine Bay where they were picked up by Captain Halsey aboard the pirate brigantine *Charles.* When the main party arrived in the capital the one-eyed king Andrian-Kirinda informed them that he was

Figure 44. The sinking of the *Degrave* (from Robert Drury's journal, 1729 edition)

conscripting them into his army to fight in the unending wars with neighboring kingdoms. This led the crew to plot an escape. In a swiftly executed uprising, they took the king, his wife, and one of his son's hostage together with a stock of arms and ammunition. They fled eastwards towards the river Mandrare dividing the Androy country from the Anosy kingdom. They hoped they would be more hospitable to Europeans. It was a desperate decision. They were heading across a waterless semi-desert towards Fort-Dauphin and its ruler King Samuel.

Pursued by an army of 2,000 warriors, the sailors liberated the king and the queen on the second day keeping only the prince as hostage. On the third day, they were tricked during a parley into releasing their last hostage, which was a fatal mistake. The Antandroy did not turn back as promised but started closing in. On the fourth day, they left at daybreak and managed to reach the river in the morning. As they waded the river and climbed up a sand hill on the other side, the Tandroy army was on their heels. The seamen kept their attackers at bay until nightfall by which time their powder and shot were exhausted. Under the cover of darkness, a group of 30 men, including Captain Drummond, slipped away, abandoning their comrades to their fate. One of these escapees was John Benbow nephew of the famous admiral after whom Stevenson named the inn in *Treasure Island.* Benbow made it back to Britain. After four years in southeast Madagascar he was taken off by a Dutch slave ship and filed a report on February 9, 1707 in Amsterdam.

The Tandroy army slaughtered all the men who stayed on the sand hill with the exception of three boys, the *Degrave*'s midshipmen including Robert Drury. Separated from his shipmates and dragged back through the woods and sand dunes to the village of the king's grandson, Maharavia, Drury began eight years (1703-1711) of captivity as the man's personal slave. Life was hard. The Tandroy people survive in this arid country by raising cattle and Drury spent years as a cattle-herder. As he grew up, he became a trusted member of the household and was sent with the others into cattle-stealing raids on neighboring villages. Nonetheless Drury was often humiliated. He wanted to escape but was in a dilemma. He had fallen in love with a fellow captive the beautiful daughter of a defeated neighboring chief. Superstition had it that slaves were bound to their masters by magic spells which killed all runaways. Both Drury and the girl were subjected to bewitchment ceremonies and the girl was too afraid to leave.

Shortly thereafter Drury met Ranoana, an envoy of the King of Fiherenana, a kingdom to the northwest of Androy where Europeans lived as free men. It was located close to the Bay of St. Augustine where English ships occasionally called for provisions. Ranoana offered to buy him but Maharavia refused. This strengthened Drury's resolve to escape. Ranoana explained to him how to get to the Bay of St. Augustine. Drury walked out of Androy alone hoping to enlist with the Fiherenana army. He had been mistakenly informed that it was close by. Not finding the troops, he took refuge in a village led by Andrian-Aferana who took him in as a kind of page. About six months later, Andrian-Aferana did not allow him to join the Fiherenana troops and brought him back to king Andrian-Kirinda in Androy. After participating in a war against the Mahafaly and now 26 year old, he fled one night and followed the itinerary that Ranoana had given him to get to the Bay of St. Augustine.

After a three weeks' journey, Drury crossed the Onilahy River. He followed its bank heading west and arrived at the Bay of St. Augustine a week later. He was warmly received by chief Andriamitranga. He lived a calm life as he waited for an English ship. He befriended an old Dutch pirate, Eglasse, who was disembarked from the corvette of the American pirate Samuel Burgess and was living there with two slaves.

However, Drury's relative safety was short-lived. The village where he lived was attacked and he was captured again, this time by the army of the Sakalava, a kingdom that controlled the western half of Madagascar. His new master, the Sakalava General Ravovy took him to Mahabo, the capital of South-Menabe, where he was introduced to King Tsimanangarivo. The king told him of another Englishman, Thornbury, who was abandoned on the coast and was awaiting another ship to return to England. Later, Drury fled his master Ravovy and took refuge with the king's oldest son, prince Ramoma, where he lived for two and a half years accumulating his own small herd of cattle.

Rescue finally arrived in 1717. The English ship *Drake* turned up to trade for slaves and his master allowed him to leave on the voyage home. In London there was not much call for his skills of spear-throwing and cattle-herding—and nobody except the slave traders needed a Malagasy interpreter. He wrote the East India Company to employ him on another expedition to Madagascar. If they couldn't find him a job, he wrote he would work for the Swedes.[99] Within a year he was aboard the *Mercury* bound for Madagascar returning to ply his new trade as a slaver. His second stay lasted two years, 1718-1720, during which he procured slaves in Fort Dauphin, Matitanana, the island of St. Marie and finally the Bay of St. Augustine before heading off for Virginia.

In the late-1720s he became a doorman at the HEIC headquarters in London. Until his death in 1735, Robert Drury could be found frequenting Old Tom's Coffee-House in Birchin Lane in central London. Here, he 'was willing to confirm those Things which to some may seem doubtful' and 'gratify any Gentleman with a further account' of anything in the book. He was buried in the churchyard of St. Clement Danes in the Strand.

At the site of the wreck of the *Degrave,* two iron cannons lie on the reef. Several others and an anchor are on the sea bed. Some years ago local divers recovered part of an English ship's bell whose style dates it to before 1750. It is now located at the Art and Archeological Museum in Antananarivo. Somewhere on the bottom lies the rest of the bell and it may well have the name of the ship or the Company's mark on it.

Figure 45. Cannon from the *Degrave* on the reef at Tanroy (photography by Tim Healy)

Figure 46. Fragments from the *Degrave* bell (courtesy of the Archeological Museum of Antananarivo)

AURORA[100] (1770)

HMS Aurora was a 5th rate, 679-ton British Royal Navy frigate armed with 32 cannons, built in 1765 at H.M. Navy Dockyard in Chatham and launched on January 13, 1766. She was later purchased by the HEIC for its "India Squadron" (also called "Bombay Marine"), a small fleet of Navy vessels used in India for the defense of both HEIC ships and Indian vessels sailing under the passport and the protection of the Company.

In 1768, in consequence of complaints of misconduct by servants of the HEIC, the directors sent three surveyors, Henry Vansittart, at the time Governor of the Bengal government for the HEIC, Luke Scrafton, and colonel Francis Forde, with authority to investigate and redress grievances. The *Aurora* weighed anchor from London on September 20, 1769 under the command of Captain Thomas Lee. Her purser was the Scottish poet William Falconer, author of "The Shipwreck, A Poem" (1762) and of the "Universal Marine Dictionary." Also on board were the astronomer reverend William Hirst, fellow of the Royal Society, and the 17-year old Robert Pitcairn, who, as a 15-

Figure 47. A 5th rate Royal Navy vessel typical for the period

year old midshipman on the *HMS Swallow*, had discovered Pitcairn Island on July 3, 1767.[101]

After an uneventful sailing down the Atlantic Ocean, *Aurora* reached the Cape of Good Hope safely. Having refreshed, she set sail for Bombay on December 29, 1769, intending to provision on the way in Anjouan Island (Comoros). Neither the *Aurora* nor its crew were ever seen or heard of again. The conjectures were that the ship either went down in a storm in the Mozambique Channel, or burned at sea, or wrecked on Madagascar's south coast. A vessel of the French East India Company having reported to have distinguished debris of a shipwreck in the south of Madagascar at around the time of the *Aurora*'s disappearance (January 1770), the latter hypothesis was given some credence. Searches were made by Sir John Lyndsay in every place she might have been shipwrecked, but to no avail.

Some large anchors and cannons, likely of British making, having been found close to Star Bank (Banc de l'Etoile) would indicate that the *Aurora* probably foundered in that area with all hands lost.

WINTERTON[102] (1792)

The 3-deck 771-ton East Indiaman *Winterton* was built at the Perry shipyards, launched at Blackwall, London on January 2, 1782 for Thomas Newte and was in service to the HEIC for 10 years. Her first two voyages (1782-1786) were commanded by Raymond Snow. Captain George Dundas took her for a third voyage (1788-1790). Captain

Figure 48. Anchor found in the South of Madagascar that could belong to the *Aurora*

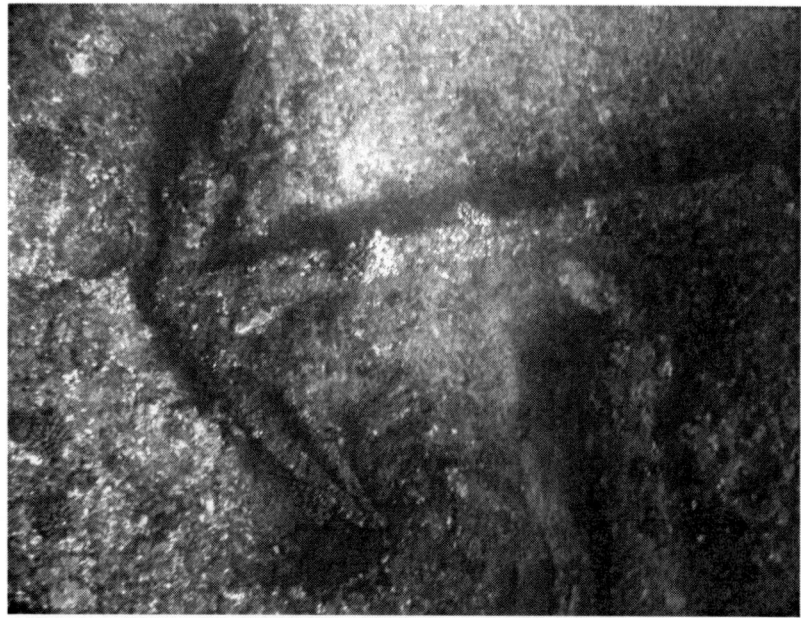

Figure 49. Anchor found in the South of Madagascar that could belong to the *Aurora*

Dundas left for her fourth voyage destined to Madras and Bengal on April 29,1792. There were about 280 passengers on board including ten women and a precious cargo worth over £100,000 (equivalent to $250 million today) that included about 300,000 silver Pieces of Eight.

After rounding the Cape of Good Hope, Captain Dundas chose to sail the Inner Passage, i.e. the Mozambique Channel as the winds were not favorable for rounding Madagascar to the East. He had planned to refresh at the Bay of St. Augustine but contrary winds prevented him from doing so. On Sunday, August 19, a southwestern wind sprang up. In the afternoon, the crew thought they saw breakers and a sloop was lowered to find out. But the Captain decided that the distant foam was caused by the far-off spouting of whales surrounding the *Winterton,* and the sloop was recalled. When retiring at ten at night, Captain Dundas said that he smelled land and that they would see it the next morning. Notwithstanding a perfect night under a new moon third officer John Dale was anxious. The light wind brought a smell of land to his nose sapping his confidence in their position. At three in the morning Captain Dundas came back on deck and noted the position on the map. He estimated they were about 60 miles from the coast and retired again. A few minutes later, the ship was jolted by a violent shock as it struck a reef.

Emergency procedures were immediately launched including throwing all the sails back and lowering the jolly-boat and yawl for

Figure 50. The *Winterton*, detail from a painting by Thomas Luny, c. 1783

Figure 51. The *Winterton*, detail from a painting by Thomas Luny, c. 1783

sounding. They braced the yards around backing the sails so that the wind right off the shore would push the ship backwards. The ship at one time got off and gained some sternway. However, confused by the darkness of the night and before she had gone sufficiently astern to admit of steerage-room orders were given for filling the sails again. The vessel was once more fatally precipitated on the reef of surrounding rocks. With the tide ebbing, they swung out the cutter along with the kedge anchor and hawser, 860 feet astern. Aboard the ship the crew tried to heave the ship off on the capstan but it failed. Then the crew cleared the upper deck, furling the sails, swinging out the longboat, the only remaining craft left aboard, striking the topgallant yards and masts and rafting the booms alongside so that they floated rather than added weight.

By the time these different operations were effected, daylight appeared on Monday, August 20. They found that they were stuck on a reef about six miles from land. Captain Dundas ascertained that they were at Point St. Felix (Ambatomifoka off Salara) 63 miles north of the Bay of St. Augustine. They also saw the mast and some other remnants of a large Portuguese ship that had wrecked there a few years earlier.[103] Because of the battering during the night, the ship was leaking badly and the soldiers earnestly manning the pumps could do little to stem the ingress of water. However, Captain Dundas had still not abandoned hope of getting her off at the next high tide that afternoon.

After breakfast, they threw as many heavy articles as they could overboard while the third mate, John Dale, carried the heavy guns well away from the ship so that she would not strike them if she managed to float again. The surf intensified with a sea breeze setting in, pounding the ship with violent waves and breaking the booms that were fastened alongside adrift which greatly impaired the means of building rafts. When high tide came between 3 and 4 in the afternoon, a final attempt was made to heave the ship off by hauling on the kedge anchor. However, she would not budge. With hindsight this saved lives as the leaks had by now gained so much that she would have foundered in deep water if she was dislodged.

The ship was lost. The next primary consideration was to organize an orderly evacuation. The masts were cut and sent overboard while securing spars as raft building material. The yawl was dispatched at sunset with second mate Nathaniel Spens and the purser to establish the safest landing place. In the evening, Captain Dundas called everyone together and explained that they would be brought to shore the next morning and were to make their way to St. Augustine Bay. The passengers retired in the round house to an uneasy night. Around midnight, while the surf was beating against the wreck with furious violence, a cry was raised that the ship's boats that were moored astern were capsizing. Ropes were thrown out but only three of the ten men who had remained in the longboat were saved from drowning.

At daybreak on the 21st, with the boats sunk, raft building began. It was done chiefly from planks as the best timbers were gone. The carpenter suggested sawing off the poop deck to use as a floating platform which the Captain seized upon. About nine that morning, the yawl returned from the shore but stood off to avoid being overset. The second mate reported that the shore was covered with rough surf in every direction and there was no calm landing place. The yawl then turned back and sailed to St. Augustine's Bay in the hope of procuring assistance. It was observed that articles thrown overboard were steadily drifting inshore, pointing to favorable wind and current conditions. Three or four rafts left the ship with about 60 people on them while others went on remaining hen-coops and empty wine chests eventually being safely washed up on the shore thanks to the rising tide.

In the afternoon, it was found that the poop could not be launched that day because of the state of the tide. John Dale and his men cut the driver boom in three lengths and with some spar brought up from the orlop deck were able to make themselves a kind of catamaran. The ship was gradually heeling over to larboard and by evening the deck lay nearly perpendicular. Between 6 and 7 in the evening, the hawser attached to the kedge anchor snapped. The ship was hove round broadside on to the rocks and took the full fury of the pounding water. Everybody on board scrambled to the starboard side which was now uppermost including people on the lower decks who were plundering the *Winterton*'s treasure of silver coins. There was a shout for all to get on the poop.

Soon after this, the ship broke apart with a dreadful crash somewhere near the fore hatchway. The forepart was carried into the sea with some people on it. All in all, there were still about 180 to 200 people on the poop and the starboard side of the ship. Dundas began to help the ladies and other passengers onto the poop which was soon washed away with 80 to 100 people on board. John Dale and the other junior officers launched their raft. A few minutes later, a massive wave shattered what was left of the ship washing off the 100 or so people still remaining on it. Forty persons, including Captain Dundas, lost their lives in the frenzied surf. Survivors clung to whatever piece of wreckage they could find. A great piece of the starboard side with cabins and bulkheads still firmly attached below became a raft for about 50-60 of the crew and passengers.

Figure 52. Surf breaking over the *Winterton* as the first rafts leave the ship. The ship's masts and rigging have been cut away and the kedge anchor continues to hold the hull broadside-on to the surf. (Source: Buchan, "A Narrative of the Loss of the *Winterton*," 1820)

Figure 53. Final moments of the *Winterton*: the poop floats free, John Dale's group cling to their raft, and the remaining crew and passengers are on the starboard side (Source: Buchan, "A Narrative of the Loss of the *Winterton*," 1820)

Most of the crew and passengers were eventually washed up, exhausted, on the beach—the first on the afternoon of the following day (Wednesday the 22nd). The last group landed on Sunday (the 26th). Some were on the poop, some on small pieces of the wreck which drifted near the shore and others rescued in canoes manned by the natives. All of the 80-100 passengers on the poop survived except for a soldier who loaded himself with silver coins before leaving the wreck then decided to swim to the shore when approaching the land. He was dragged to his death by the weight of his booty. The whole number who were drowned was recorded as 48 in Mr. Dale's journal. The survivors set out on their march towards St. Augustine Bay in several groups. In a few days they arrived at Tulear, the residence of the King of Baba. He treated them with kindness, generosity and universal benevolence during their stay.

John Dale, the third mate and second senior surviving officer organized a rescue mission in the *Winterton*'s one remaining yawl. He rigged up the yawl and set off on September 13, 1792 with the fourth mate, Mr. Wilton, a Portuguese, Mr. de Souza, and four seamen. They left without proper charts and at the mercy of the currents to fetch help and procure a vessel in Mozambique Island some 600 miles to the northnorthwest. With luck, he hoped to return with a ship in a few weeks. However, once they reached the African coast contrary winds and currents pushed them south and they took relief at the Portuguese settlement in Sofala.

They attempted an ill-fated navigation with the leaking Sofala settlement boat but after two weeks were forced to return to Sofala. On October 29, they set off on a torturous trek into the hinterland of Mozambique. They reached Sena on the southern bank of the Zambezi on December 6. Seriously ill and exhausted, three seamen died there. The three survivors left Sena in small canoes on January 19, 1793 reaching the delta port of Quelimane in a few days. Leaving one man behind, Dale and de Souza boarded a small sloop bound for Mozambique on January 31, arriving two weeks later. Dale was able to procure a small 160-ton snow and sailed for Madagascar alone on March 1. He arrived in St. Augustine Bay on March 24, more than 6 months after his departure.

Dale was confronted by a miserable sight. Climate, living conditions and fevers had reduced the party by half to one hundred and thirty and most survivors were sick and emaciated. Dale evacuated all of them to Mozambique in April 1793 except for one sailor who fell in love with a young native and asked to stay. After receiving some medical treatment in Mozambique, 85 survivors of the *Winterton* embarked on June 10 in a private vessel, the 250-ton brig *Joachim*, which Mr. Dale had freighted for Madras. After picking up supplies in Anjouan and having covered almost half the distance to Ceylon they were captured by the French privateer, *Le Mutin,* Captain Jean Mallet. Dale and twenty-two seamen and soldiers were taken into the privateer. Two French officers and 10 seamen came aboard the *Joachim* to guard the remainder and steer for Mauritius.

The privateer proceeded on her cruise and off Tuticorin she engaged the Dutch Indiaman *Ceylon.* However, after about a quarter of an hour the *Mutin* was captured. The British seamen were liberated. Dale and his men reached Madras on August 20, 1793, twelve months to the day of the shipwreck. Dale started his return journey to England aboard the *Scorpion.* Off Brest, the *Scorpion* was captured on January 12, 1794 by a squadron of French ships sailing to the Chesapeake Bay. After being freed, he went to Philadelphia. He finally returned to England on July 11, 1794 two years and two months after he had sailed. The rest of the party arrived in Mauritius on August 17, 1793 but were stranded by the embargo on shipping enforced by the French to prevent a leak of information on their privateer activities. In December 1793, they were brought to Madras aboard the American ship *Henry.*

The wreck was located and explored in 1985 by the marine archaeologist Dr. Robert Sténuit, founder of the Groupe de Recherche Archéologique Sous-Marine Post-Médiévale. This author has dived on the site several times.

REFERENCES

[88] Sources used for the introductory paragraphs are British East India Company, Wikipedia at http://en.wikipedia.org/wiki/British_East_India_Company; Kerr, R., "A General History and Collection of Voyages and Travels Arranged in Systematic Order: Forming a Complete History of the Origin and Progress of Navigation, Discovery, and Commerce, by Sea and Land, from the Earliest Ages to the Present Time," 1813; Hakluyt Society, "Voyage to the East Indies," 1877; John Keay, "The Honourable Company—A History of the English East India Company," HarperCollins, London, 1991; Grandidier et al., op. cit., t. I, pp. 160-163, 276-280, and 402-407.

[89] The English seaman Henry May aboard the *Bonaventura* was transferred by his Captain, James Lancaster, to a French vessel under the command of M. de la Barbotière bound for Europe, to report on British ships and British prospects

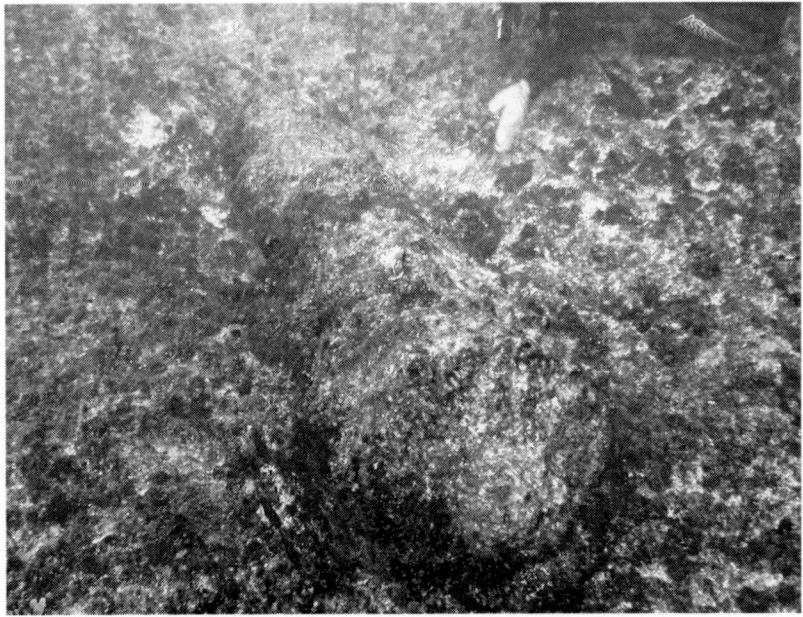

Figure 54. Cannon from the *Winterton*

Figure 55. Anchor and ballast

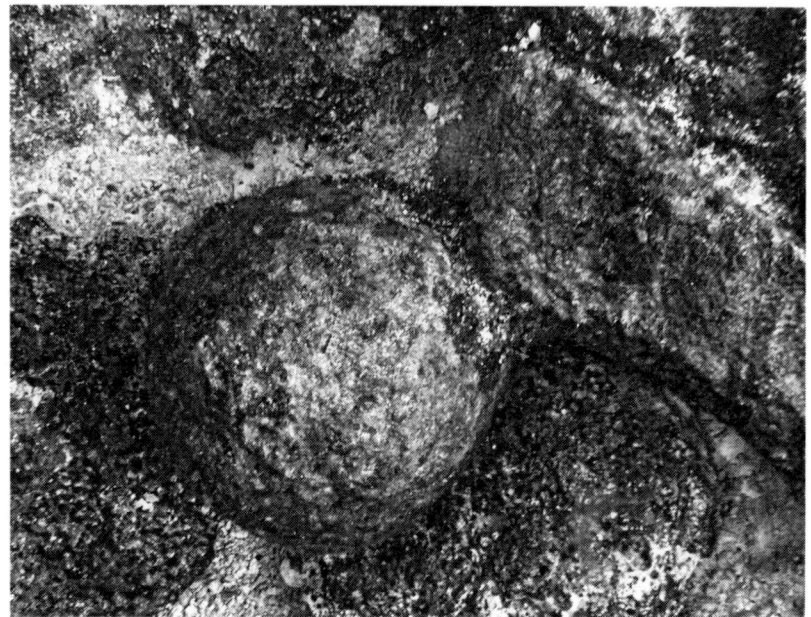

Figure 56. Cannonball

Figure 57. Other cannons

Figure 58. Kedge anchor placed 860 feet astern

Figure 59. Lion statuette

Figure 60. Copper lining and Piece of Eight

Figure 61. Shoe buckle

for the region. He related the tragic story in Kerr, R., op. cit., vol. VIII, pp. 17-18 and 32.

[90] Hakluyt, R., "The principal Navigations, Voyages, Traffick, Discoveries of the English," London, 1599, t. II, p. 286; Hackluyt, R., "Collection Early Voyages," t. II, 1810, pp. 587-588; Brooke: History of St. Helena, at http://www.bweaver.nom.sh/brooke/brooke_ch2.html.

[91] It is possible that the 453-ton vessel *Samaritan,* purchased by the HEIC in January 1614 and outfitted for a voyage to India later that year, foundered off Madagascar in 1615 during the outward voyage. Further investigations are needed to ascertain this.

[92] Source: Letter written by Captain William Freke at Bombay Castle, December 8, 1689, British Library, India Office Library and records, No. 5690, pp. 93-98.

[93] Hindu traders especially from Gujarat.

[94] Captain Abraham Samuel, a mulatto pirate from Martinique, wrecked his ship *John and Rebecca* in Fort Dauphin in 1697 (see below, section on Pirates). The old queen recognized him as her son, whom she had conceived with a Frenchman long before and who had left with his father as a child. Captain Samuel, having become the sole heir to the throne, fought and subdued the local ruler and became the King of the Anosy region. However, the massacre of the remaining small French contingent in Fort Dauphin had taken place a quarter of a century before, in 1674, when Governor de la Bretèche departed for Surat, definitively giving up the remnants of a colony enfeebled by daily skirmishes with the local population and internal dissent (see below under *La Dunkerquoise*).

[95] Drury, R., "Madagascar: or Robert Drury's Journal during fifteen years captivity on that island," London: W. Meadows, 1729; reprinted in 1807 by Stodart and Craggs, Hull. See also Desperthes, "Perte du Vaisseau de la Compagnie des Indes, Le Degrave, sur la Côte de Madagascar, en 1701, et aventures de Robert Drury," Ed. Eyriès, 1821, t. II, pp. 138-184.

[96] Secord, A.W., "Robert Drury's Journal and Other Studies," Urbana: Illinois University Press, 1962.

[97] See Pearson, M.P. and Godden, K., "In Search of the Red Slave: shipwreck and captivity in Madagascar," Sutton, 2002; Pearson, M.P., "Shipwreck into Slavery," British Archaeology Journal, Issue 67, October 2002 at http://www.britarch.ac.uk/ba/ba67/feat2.shtml; Sheffield University: "Bibliography for Robert Drury, his life and journal" at http://www.shef.ac.uk/archaeology/research/madagascar/robert_drury

[98] After wrecking their ship *Speaker* in Mauritius on the return from a run in the Red Sea and the coast of Malabar, and leaving behind the Moor and Banian crew that the *Degrave* rescued two years later, the pirates under Captain John Bowen set down in March 1702 in Matitana on the eastern coast of Madagascar and built a fort and town. Later that year, a ship, the *Speedy Return,* set into Matitana along with a brigantine, *Content,* having just taken on slaves in St.

Mary's. Bowen and some others snuck aboard and took possession of both ships, leaving Captains Drummond and Stewart and some of their men ashore. After watering and provisioning both ships, the pirates set sail. The brigantine ran on some rocks off the west coast of Madagascar and only caught up with the *Speedy Return* at Augustine Bay. Here the pirates abandoned and burned the *Content*. Having heard that the pirate Captain Thomas Howard had departed soon before aboard the *Prosperous,* they went in search of him in an effort to ally. Captains Drummond and Stewart and their men used the longboat to try to reach St. Augustine Bay. Having passed to the southward of Fort Dauphin, the wind shifted and blew so hard that they could not carry sail, so they drove on shore within three or four leagues of the spot where the *Degrave* wrecked. (See: Captain Johnson, Ch., "General History of the Robberies and Murders of the most notorious Pyrates," London: J.M. Dent & Sons, 1724, vol. 2, pp. 51-52 and 58-59).

[99] East India Company. *Court Minutes,* volume 53 (1730), 444.

[100] Sources: Phillips, M., "Ships of the Old Navy" at http://www.ageofnelson.org/MichaelPhillips/info.php?ref=0255; Lettens, J., "HMS Aurora" at http://www.wrecksite.eu/wreck.aspx?16210; Edwards, R. "Robert Pitcairn," 2004, at http://www.findagrave.com/cgi-bin/fg.cgi?page=gr&GRid=9288829

[101] On which Fletcher Christian and the mutineers of the *Bounty* settled in 1790.

[102] Sources: [Buchan, George], "Narrative of the loss of the Winterton East Indiaman, wrecked on the coast of Madagascar in 1792; and of the sufferings connected with that event. To which is subjoined, a short account of the natives of Madagascar, with suggestions as to their civilization," by a passenger in the ship, Edinburgh, 1820; Dale, John, "A Narrative of the Loss of the Winterton," C. Withington for B. Farrington, c. 1796; Anthony J., "Catalogue of East India Company Ships' Journals and Logs 1600-1834," British Library, 1999; Hackman, Rowan, "Ships of the East India Company," World Ship Society, 2001; Hood, Jean, "Marked for Misfortune: An Epic Tale of Human Endeavour and Survival in the Age of Sail," London: Conway Maritime Press, 2003; Sutton, Jean, "Lords of the East: The East India Company and its Ships 1600-1874," London: Conway Maritime Press, 2000.

[103] The *Nossa Senhora do Monte do Carmo* (see above).

4

The French East India Company

The French were well aware of all the riches that the Portuguese were collecting in the Indies. The French Kings reigning in the 16th century, François I, Charles IX, and Henri III, exhorted their subjects to undertake long-distance sea voyages. However, little came of it because of general French disinterest in the early discoveries in the Indies. Also, religious wars against Calvinists were bankrupting the country at the time.[104]

The first French ship recorded to have reached Madagascar and the first French wreckage, was in 1527.[105] Three ships left Dieppe in 1526 travelling to China and the Moluccas via the Magellan Strait. This expedition was financed by a bank in Florence with the support of the French King François I. It was set up by Normandy entrepreneur Jean Ango and the Verrazano brothers. The ships left Dieppe on June 15, 1526 but were separated at sea in a storm. The Verrazano brothers explored Brazil while the third ship, commanded by Pierre Caunay, rounded the Cape of Good Hope and the south of Madagascar and reached Sumatra in the summer of 1527. Things did not go to well in Aceh. After the pilot and several crew members were murdered, Captain Caunay decided to head home. On the way back, he anchored in a bay on the southeast coast of Madagascar but was thrown on the rocks while trying to leave the bay. Twelve of the survivors used the sloop to sail to Mozambique. They arrived on July 18, 1528, were taken prisoners and told their tale which was later reported by the Governor of Mozambique. Of the sailors who stayed in Madagascar one was picked up in 1531 by Diogo da Fonseca who had anchored the *Santa Maria* in the harbor of Ranofotsy.[106]

In late 1528, Jean Ango put together another expedition to Sumatra and the Moluccas with the 200-ton *Pensée* and the 120-ton *Sacre* commanded by the brothers Jean and Raoul Parmentier.[107] Madagascar was sighted on July 24, 1529 and both ships landed at the mouth of the Manambolo River on Madagascar's west coast four days later. A small team sent out to fetch fresh water was attacked by local inhabitants and three of the Frenchmen were killed. Disgruntled by this hostile reception, the ships set sail the next day. They explored the Barren Islands and then sailed to the Comoros and Sumatra. Captain Jean Parmentier died in Sumatra on December 3, 1529. His brother Raoul died from fevers on the way back. The two vessels returned to Dieppe in July 1530 in a sorry state carrying only 30 barrels of pepper as its sole commercial yield. However, they brought back many strange things for those times including two orangutans! A sculptor later chiseled that pair's image on a wall of a chapel of the church Saint-Jacques in Dieppe.

There were no further voyages of French ships to the East Indies in the 16th century. This was due to lack of official and financial support by the crown and because the private investors bankrolling these early expeditions lost confidence in the financial interest of trading in the East Indies controlled by the Portuguese.

The next attempt by the French to sail to India was in 1601. Merchants from St. Malo, Vitré and Laval commissioned two ships, the 400-ton *Le Croissant* and the 200-ton *Le Corbin*.[108] The ships left St. Malo on May 18, 1601. They were caught in a storm in February 1602 off Madagascar forcing them to take relief in the Bay of St. Augustine. They were joined by the Dutch ship *Ram* that was damaged in the same storm. They spent three months there. During that time 40 crew members died and were buried there. They left the bay in May 1602 and sailed to India via the Comoros Islands. The *Corbin* foundered in the Maldives but the *Croissant* reached Aceh in late 1602. She was abandoned in the Atlantic in May 1603 during the return voyage. Her cargo was given to the Flemish crew who came to her rescue.

Contrary to the Portuguese, Dutch and British whose goal was to trade with India, the French focused on setting up a colony on Madagascar's east coast. A series of ephemeral trading companies were created. Each was granted exclusive rights for trading in the East. In 1604, a renegade Dutch named Pieter Lintgens convinced King Henri IV to set up a French India Company with a trade monopoly for 15 years. Most of the initial shareholders were Flemish. Lintgens died shortly afterwards and the management of the Company was trusted to Girard de Roy, a Flemish who had already traveled to India. In 1611, King Louis XIII granted exclusive trading rights to the *Compagnie de Montmorency pour les Indes Orientales.* This generated little commercial activity. Some traders from Caen then tried to take over the privilege. In July 1615, the King sided with the old Company and merged the two parties by granting exclusive rights to the *Compagnie des Moluques* for 18 years. Once more, nothing much came of it largely because the Thirty Years war erupted in 1618. This religious conflict ravaged most of Europe. It brought defection by traders of St. Malo and Rouen who each wanted their own companies.

Cardinal de Richelieu came to power in 1624. He signed the Treaty of Compiègne with Holland. This established freedom of trade with the East and West Indies and rekindled French interest in trading with the east. In 1635, Captain Gilles de Régimont from Dieppe created his own trading company with other merchants. He undertook successful voyages to India and Madagascar. In 1642, shortly before his death, Cardinal de Richelieu granted the *Compagnie d'Orient* a monopoly for trade with Madagascar and the neighboring islands for 10 years. This group was composed of 24 partners from Dieppe headed by Captain Rigault. This company soon foundered because of lack of funds. This company dispatched the *Saint Louis* (see below) in 1642 with a small party of Frenchmen under the command of Mr. Jacques Pronis to establish a settlement at the Bay of St. Luce.

A year later, this swampy mangrove site was abandoned in favor of a healthier peninsula 35 kilometers to the south. A fort was built and named after the Dauphin (later Louis XIV). Because of internal dissent, in February 1646 the settlers rebelled against Mr. Pronis and imprisoned him. He was reinstated in July 1646 by Captain Le Bourg of the ship *Saint-Laurent* but a few months later became the victim of a second mutiny. He suppressed it and exiled 12 of the mutineers to Reunion Island (then called Mascareignes). They became the first European settlers on that island. After hearing of this sorry state of affairs, the Company recalled Mr. Pronis. In May 1648 they again dispatched the *Saint-Laurent* with a party of 80 and Mr. Etienne de Flacourt as Governor of the settlement at Fort Dauphin. Mr. Flacourt remained in Fort Dauphin for over six years.[109]

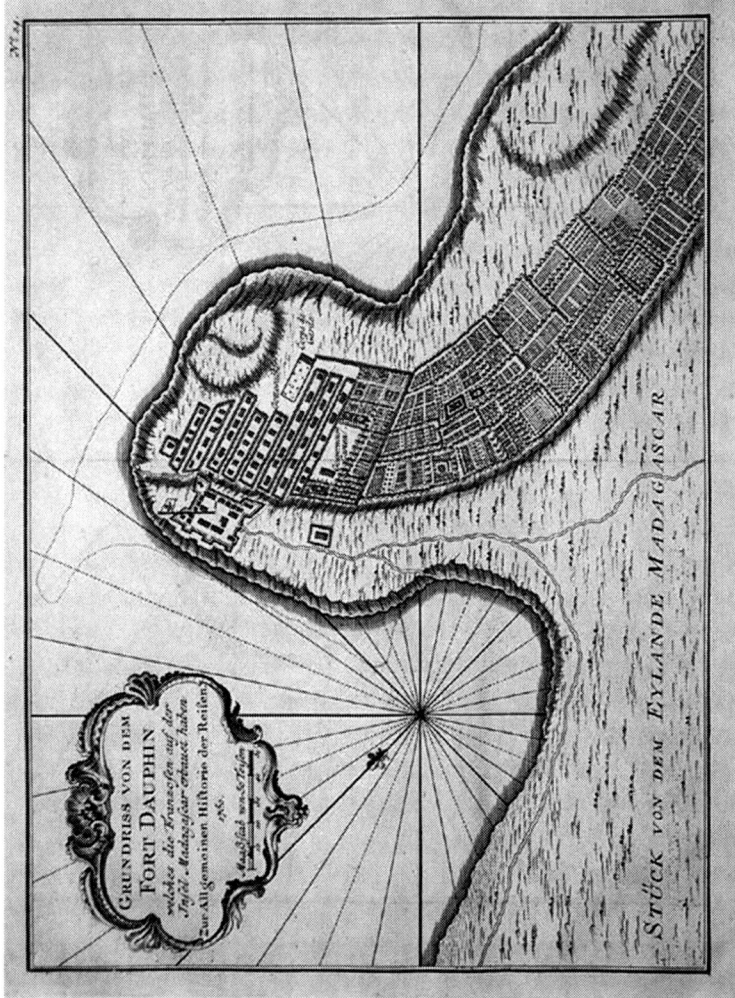

Figure 62. Map of Fort Daupin (Based on an etch by Jacques Nicolas Bellin, 1750)

De Richelieu's successor as Minister of Commerce, Marshal Duke de la Meilleraye, eschewed the *Compagnie d'Orient*. With the support of Mr. Fouquet and other merchants he created his own *Compagnie de la Meilleraye* with the blessing of the King. The company sent several fleets to Madagascar starting with the *Saint-Georges* and l'*Ours* in 1654 under the command of Captain la Forest des Royers. That company also folded when a number of disastrous expeditions bankrupted it.

To improve France's inferior role in the trade with India, Jean-Baptiste Colbert planned to set up a French East India Company as a commercial enterprise to compete with the British and Dutch East India companies. King Louis XIV granted a charter to the *Compagnie française des Indes Orientales* in 1664. Her first Director, François Caron, who had spent 30 years working for the Dutch East India Company[110] sailed to Madagascar in 1665. While the Company failed to set up a colony on Madagascar, it established ports on the islands of Bourbon (Reunion) and Île-de-France (Mauritius). By 1719, it had established itself in India, but was near bankruptcy. It was combined under John Law with other French trading companies to form the *Compagnie perpétuelle des Indes*. It gained its independence again in 1723. With the decline of the Mughal Empire, the French decided to intervene in Indian political affairs to protect their interests by forging alliances with local rulers in south India. From 1741 onward, the French, under Joseph François Dupleix, pursued an aggressive policy against both the Indians and the English until their ultimate defeat by Robert Clive. The Company was not able to maintain itself financially and was abolished in 1769. In 1785, another company was created, the *Compagnie des Indes orientales et de la Chine* but it foundered in the turmoil surrounding the French Revolution.

Given the strong French presence in the southwestern Indian Ocean, including the various attempts to colonize Madagascar and the establishment of neighboring settlements in Reunion and Mauritius, numerous French ships wrecked in Madagascar.

SAINT ALEXIS[111] (1641)

The *Saint Alexis,* a 300-ton flute armed with 14 cannons and 97 men on board, left Dieppe on January 15, 1638 under the command of Captain Alonse Goubert. She was bound for the Red Sea via Mauritius. She was accompanied by the *Marguerite,* under Captain Grégoire Digart. The *Saint Alexis* was carrying an 80-ton yacht that was to be reassembled in Madagascar for running the Red Sea. She took relief in the southeast of Madagascar in May of that year and then proceeded to Mauritius to load a cargo of ebony wood, anchoring in the southeast port on June 8, 1638. The next day, Cornelis Simonsz Gooyer, the first Governor of Mauritius,[112] paid a visit and inquired as to the purpose of her stay. When Captain Goubert stated that they only came for refreshments the Governor allowed the crew to disembark to go hunting and fishing.

The next day, the *Saint Alexis* rounded Mauritius and anchored in the northeast (the actual Saint Louis) that was defended by six Dutch soldiers. The Dutch saw that the French were setting up tents onshore and offloaded carts, obviously to cut and harvest a cargo of ebony wood. But there was little they could do as their instructions did not mention that such activity needed to be ruled out. The cutting and loading took several months. On December 12, 1638 the *Saint Alexis* was ready to sail and asked the Governor's permission to leave.

Having received no order to the contrary, Governor Gooyer let her go on December 20, giving Captain Goubert a letter for the VOC to be mailed from Dieppe to Amsterdam. The *Saint Alexis,* however, sailed to St. Luce Bay, touching on Rodrigues and Reunion islands along the way but did not go back to Dieppe as intended.

The *Saint Alexis* returned to Mauritius in 1640 anchoring in the harbor of Saint Louis on June 10. Captain Goubert carried orders to take possession of Mauritius for the French king, offload some of his crew to set up a settlement and get more ebony wood. The new Dutch Governor of Mauritius, Adriaan van der Stel, prevented him from doing so. Captain Goubert protested in writing and set sail for Madagascar on June 21, 1640.

Back in St. Luce in July 1640, they started reassembling the yacht. Some dissent arose between the Captain and the Master, Jacques Soulas. The Master wanted to load the yacht with ebony wood and both ships return to France. The Captain argued for a run to the Red Sea in search of a good capture. After about six months, several crew members were stricken by a contagious tropical disease and died. It was decided to look for a healthier abode. The healthy boarded the *Saint Alexis* with the supplies while the sick sailed with the yacht. The two ships sailed to Saint Claire (Itaperina) eight leagues to the south arriving there in March 1641. When the contagion ended, their numbers were reduced by about half with about 50 crew members remaining. Furthermore, it was found that the ship was seriously affected by shipworm and was not seaworthy enough for the return trip to Europe. The crew moved the cargo, cannons, ammunition and lifting gear to a warehouse and beached the ship in the Itaperina cove.[113]

The crew returned to St. Claire. In early 1642, Governor van der Stel aboard the *Eendracht* anchored there having secured slaves in the Bay of Antongil accompanied by the yacht *Klein Mauritius*. They took 17 of the crew members of the *Saint Alexis* back to Mauritius. The yacht was reinforced, supplies secured and the Master, Jacques Soulas, sailed in March 1642 for the return trip to Dieppe with 20 men. Meanwhile, the *Marguerite* had returned from a run in the Red Sea. She took most of the cargo and Captain Goubert for the return voyage to Dieppe. Seven sailors were left behind to guard the remaining cargo[114] and lived at the Bay of St. Claire for about two years before joining Mr. Pronis and his men in St. Luce.

LE SAINT-LOUIS[115] (1644)

The *Compagnie d'Orient* created by Richelieu in 1642 sent the *Saint-Louis,* with Captain Coquet, to Madagascar in March 1643 to transport 12 men headed by Messrs. Pronis and Foucquembourg. They were charged with establishing the first settlement in the Bay of Saint Luce in Madagascar. After an uneventful voyage she arrived around September 1643.

She first sailed to Bourbon (Réunion) and Rodrigues Island which they claimed for France. To their great surprise, they found the seven marooned French sailors from the *Saint Alexis* that had wrecked in the Itaperina cove in 1641. In 1644, Captain Coquet sailed the *Saint-Louis* to the Island of Sainte Marie and the Bay of Antongil also claiming them for France. He then loaded a cargo of ebony wood in Matitanana and in the Anosy region. While sailing back, the *Saint-Louis* was thrown against the rocks by a violent wind and was lost with Captain Coquet perishing in the wreckage. The surviving crew landed her in the Bay of Ranofotsi (Gallions Bay) and sold the cargo and the ammunition to the local population.

L'ARMAND, LE SAINT-GEORGES AND LA DUCHESSE[116] (1656-57)

In late 1655, the Minister of Commerce Marshal Duke de la Meilleraye sent a fleet of four ships from La Rochelle to Madagascar with 800 men, soldiers and crew combined. The admiral ship *La Duchesse,* Captain de la Roche-Saint-André, *Le Maréchal,* Captain Colon, *l'Armand,* Captain Richmond, and the *Saint-Georges,* Captain Labriants. The fleet put in at Table Bay between March 24 and April 2, 1656. It proceeded to the Bay of Antongil where it left some sailors to set up a small settlement. Then it went to the Island of St. Marie, arriving in late May 1656. The *Maréchal* scoured the Red Sea leaving the other ships behind in St. Marie where the crew had some brushes with the Malagasy.

A small expedition sailed to the mainland with the boats. They went to Foulpointe to look for crystal rocks. They found the mine and asked the local population to help them dislodge the crystal rocks and carry them to the boats anchored in the river. As it was rice harvesting season the Malagasy refused. The French tried to compel them but the Malagasy rebelled and killed part of the small expedition. On St. Marie, meanwhile, food was short and the local population did not trade. Skirmishes ensued and the Malagasy eventually poisoned the water well that the crew had dug. Because of this and the impact of tropical illnesses about 200 Frenchmen died.

After capturing a small Turkish vessel and her cargo of ambergris the *Maréchal* returned to St. Marie in December 1656. She found the *Armand* and the *Saint-Georges* very much worm eaten. Also, having lost most of the crew to sail them both were declared unseaworthy and beached. In January 1657, the *Duchesse* sailed to Fort Dauphin arriving two weeks later. On board were Mr. Rivaux, sent by de la Meilleraye to take over the command of the colony and the 10 remaining survivors of the *Tulp* wreckage. The *Maréchal* joined soon thereafter, having stayed behind to pick up the artillery and useful debris of the two abandoned ships. The *Duchesse* was also declared unseaworthy and beached in Fort Dauphin. Her remnants were used to furbish the Governor's residence there. The *Maréchal* left Fort Dauphin on February 19, 1657 bringing four of the survivors of the *Tulp* on board. She arrived at the Dutch Cape colony on March 31, 1657 and returned to France.

TAUREAU (1666) AND AIGLE-BLANC (1667)[117]

The first fleet of Colbert's French East India Company set up in 1664 was commissioned in Brest a year later. The fleet transported 288 passengers in total and consisted of four ships: the admiral ship, the 250-ton frigate *Saint-Paul* mounted with 32 guns,[118] commanded by Captain Véron and having on board Pierre de Beausse, who was to take possession of Madagascar in the name of the King for the new East India Company; the 300-ton *Vierge-de-Bon-Port* mounted with 30 guns, Captain Truchot de La Chesnaye; the 250-ton flute *Taureau* armed with 22 cannons, Captain Kergadiou; and the small 70-ton frigate l'*Aigle Blanc,* Captain de la Clocheterie (later relieved and replaced by Girardin). These ships, bought by the Company, had seafaring experience with the *Vierge-de-Bon-Port* having done several crossings of the Atlantic and the *Taureau* having crossed the Atlantic in 1658 taking Charles and Jean Allaire with a group of settlers to Canada.

The fleet sailed from Brest for Madagascar on March 1, 1665. The voyage was uneventful except for one of *Taureau*'s boats which capsized during the refreshment in April in Cape Verde. They lost 13 lives including a priest.[119] Aboard the *Aigle-Blanc* a fight erupted between the Captain and most of the crew who belonged to the Reformed Church and Catholic passengers led by the missionary priest Bourrot. The fleet rounded the Cape of Good Hope on June 3, 1665 but was separated in a storm lasting from June 12 to 15 which slowed down their progress. The agreed upon rendezvous place was Bourbon (Reunion) island. But the *Saint-Paul* headed straight for Fort Dauphin arriving on July 10 where de Beausse officially took possession of Madagascar. On the same day, the *Taureau* anchored in St. Paul (Reunion) and disembarked about twenty settlers on the island commanded by Etienne Regnault who organized the first permanent French settlement in Bourbon. She was joined a few days later by the *Vierge-de-Bon-Port.* On July 20, the *Aigle-Blanc* arrived in St. Gilles and joined the other two ships the next day. Not seeing the *Saint-Paul* they reckoned that she had gone straight for Madagascar.

On August 8, the *Aigle-Blanc* sailed for Ghalemboule (Foulpointe, on the east coast) to drop off about 20 passengers led by François Martin who was to set up a settlement there. They were to join Mr. Belleville, one of the sailors of the *Saint-Georges,* who had lived there a few years and had married a Malagasy. They left the *Aigle-Blanc* in the harbor of St. Marie, met Belleville and went to Foulpointe with the boats. After taking on rice in Foulpointe to provision Fort Dauphin the *Aigle-Blanc* weighed anchor on October 2 arriving in Fort Dauphin on November 3. Meanwhile, the *Taureau* and the *Vierge-de-Bon-Port* had sailed from St. Paul on August 10. Because of the negligence of the pilots they missed the entrance to

Fort Dauphin bay and anchored first at Gallions Bay. Then they headed for Fort Dauphin and arrived in early September 1665. On September 17, the *Saint-Paul* and the *Vierge-de-Bon-Port* weighed anchor bound for Socotra and Mocha. They provisioned in Foulpointe and Antongil and proceeded in November. To everyone's surprise, the new Governor de Beausse died in Fort Dauphin from incurable fevers on December 14, 1665. He named Mr. Montaubon as his successor to handle a most delicate situation.

The year 1666 started in an ominous fashion for the fleet. The *Saint-Paul* encountered contrary winds after leaving Madagascar, then had problems passing the equator because of slack winds. Short on water and supplies, Captain Véron, who was never enthused by the trip to the Red Sea and Persia in the first place decided to sail back to Madagascar. On the way back, the *Saint-Paul* stuck a reef a little north of the Bay of Antongil. The crew dislodged her and she proceeded to St. Marie Island arriving in April 1666. There, Captain Véron and 20 members of the crew died. The pilot Cornuël became her new Captain. She sailed back to Fort Dauphin on August 27, 1666 and in February 1667 returned to France.

In the meantime, the *Vierge-de-Bon-Port* proceeded to Persia and her trade was successful. Overloaded with merchandise and laden with treasures she returned to Fort Dauphin for provisions and then set sail on February 20, 1666 for le Havre. On July 9, 1666, barely a day from France and safety, she was attacked and sunk by British privateers on the ship *Orange* off Guernsey. The surviving crew were taken prisoners aboard the pirate ship and brought to England. Captain Truchot de La Chesnaye died in early August while in captivity at the island of Wight. This wiped out the only positive commercial result of the first expedition of the new French East India Company.[120]

Meanwhile, the *Taureau* and l'*Aigle Blanc* sailed back and forth in late 1665 and early 1666 between Fort Dauphin, Ghalemboule and St. Marie to buy rice. However there was none. They sailed to Antongil to build a small fort and drop off 19 Frenchmen instructed to set up a small settlement there. On February 2, 1666 the *Taureau*'s Captain Kergadiou died from violent fevers and diarrhea. Later that year she lost all her officers and pilots to illness. Both ships continued the coastal trade to supply Fort Dauphin. In May 1666, the hooker *Saint Louis,* Captain de la Vigne arrived from France in Fort Dauphin. She had been dispatched with the hooker *Saint-Jacques* and 50 settlers in July 1665. However, the *Saint-Jacques* got carried away in the Atlan-

Figure 63. Capture of the *Vierge-de-Bon-Port*

tic and arrived in Brazil where she later met the second fleet (see below).

On July 11, 1666, while leaving Antongil with a shipment of rice, the *Taureau* met a strong wind at the entrance of the bay. She was improperly rigged, broke her bowsprit and became hard to steer. A westerly wind and strong currents prevented her reentry into the bay and pushed her northward along the coast. She shouldered the coast until she hit rocks south of the bay of Vohemar and immediately foundered. Nearly all crew and passengers could reach the nearby shore, save for 3 or 4 who drowned. One was Father Cuveron, the head of the missionaries in Madagascar, who, in an act of charity, returned to the wreck after reaching shore to fetch some sick passengers on board. The survivors split in small groups to maximize their chances of finding supplies and marched on a southward course while staying close to shore. After several days of exhaustion and wretchedness, during which another 6 or 7 died, the bulk of the crew and passengers arrived at the small settlement in Antongil. They were transported in batches in small sloops to the settlement in Ghalemboule. From there they departed on November 3, 1666 for Fort Dauphin aboard the hooker *Saint Louis* that had come in for supplies.

After the *Saint-Paul* returned to France in February 1667 the only vessel left to supply the colony in Fort Dauphin was the small frigate l'*Aigle Blanc*. She was used for coastal trade but foundered in a storm in mid-1667 close to Fort Dauphin.

PETIT SAINT-JEAN (1670) AND SAINT-LUC (1671)[121]

Having received no news of the first expedition due to the capture of the *Vierge-de-Bon-Port,* the second fleet of Colbert's East India Company left La Rochelle on March 14, 1666. It consisted of ten ships: the 600-ton *Saint-Jean,*[122] 36 cannons, admiral ship, the 600-ton *Marie,* 36 cannons, the 350-ton *Terron,* 24 cannons, the 300-ton *Saint-Charles,* 24 cannons, the 200-ton *Mazarine,* 24 cannons, the 200-ton *Duchesse,* 24 cannons, and four 95-ton hookers mounted with 4 cannons *Saint-Denis, Saint-Luc,* petit *Saint-Jean,* and *Saint-Robert.* The fleet was escorted for one week in European waters by four navy ships—the *Rubis,* the *Beaufort,* the *Mercoeur* and the *Infante*—to protect it from attacks by British ships with who France was at war. The fleet was commanded by François de Lopis, marquis de Mondevergue. It included the Company Directors Caron and de Faye and totaled 1,589 people; 421 officers and crew, 212 infantry officers and soldiers and 956 merchants, settlers and craftsmen. This included 32 women, several children and 10 priests.

The fleet was forced to put in at the Canary Islands to repair the *Terron* that was taking too much water. They remained there for six weeks and purchased excellent Spanish wine and another small frigate which they renamed la *Paix*. After leaving the Canaries the fleet nearly got into trouble along the Coast of Guinea due to a navigation error. Thunder caused an extraordinary incident aboard the *Saint-Charles*. A first lightning hit the poop throwing the writer on the main deck. A second jolt obliterated the main-topsail mast, bored the main mast right down to the goose-neck and caused quite a bit of damage on the main deck. Finally, an epidemic broke out aboard the *Marie* close to the equator claiming 95 lives among the 335 crew and passengers.

Slack and contrary winds held the fleet close to the equator for several days. Short of supplies, they decided to sail to Brazil. They anchored in Pernambouc (Recife) on July 25. There they met the hooker *Saint-Jacques* that had become separated from the *Saint Louis* the year before and purchased a small vessel named *Saumacque*. The fleet departed from Recife on November 2 refreshing at Table Bay between December 12, 1666 and January 7, 1667. They were warmly received by the Dutch Governor. They also explored Saldanha Bay, planted a pillar carved with the royal French coat of arms and claimed it for France. The Dutch governor had the pillar removed a few days later.

Departing the Cape, orders were to sail to Bourbon Island. The fleet got separated in a storm east of the Cape around mid-January but most ships made it to the Bay of St. Paul (Bourbon) between February 22 and 27, 1667. The crew and passengers had suffered a lot during the long passage: about 400 men had died and 27 of the women; 200 were ill and disembarked on Bourbon to be nursed to better health. After a few days at anchor they were surprised by a cyclone. While some of the officers, crew and passengers were still ashore the crew aboard cut the hawsers and moved to the open sea. They set course for Fort Dauphin where nine ships arrived one after the other in early March about a year after having left France. The *Saint-Luc* arrived on March 3rd and the *Saint-Jean* on March 10. The three other ships, having been lost along the way, came in later. The hooker

Saint-Jean which was separated from the fleet after leaving Brazil anchored on March 20. The hooker *Saint-Denis,* which was separated in Tenerife and guided by an ignorant pilot went all the way to Socotra. Then they turned back south along the Mozambique coast before reaching Fort Dauphin about a year later on March 1, 1668.

Colbert's instructions were for the *Mazarine, Duchesse, Saint-Jean* and *Saint-Luc* to proceed right away to Ghalemboule (Foulpointe or Mahavelona) to buttress the colony there before sailing to India. The other ships of the fleet were to proceed to India immediately. However, the lack of seaworthiness of the *Terron, Saint-Charles* and *Duchesse* and the desolate state of the colony at Fort Dauphin discombobulated these plans. Mondevergue decided to restructure it before proceeding to India. Food was scarce in Fort Dauphin and the arrival of Montdevergue with about 1,000 men complicated things further. One of the hookers was sent in July 1667 to Ghalemboule to procure rice and relieve Mr. Martin, the commander of the settlement who was to sail to India with the fleet. He was to be replaced by Mr. Saladin. In September 1667 it was planned to send the *Marie* back to France to inform the company of the state of affairs in Madagascar but her departure was delayed because of lack of provisions; eventually the hooker *Saint-Robert* was sent on the return voyage in March 1668. Her mission was to announce the failure of the attempt to colonize Madagascar.

Only after the flute the *Couronne* arrived from France in late August 1667 did a first convoy, consisting of the *Couronne,* the *Saint-Jean* and the hooker *Saint-Louis* leave for Surat in October 1667. The three ships came back to Fort Dauphin in June of the next year. They left Mr. Caron as Governor of Surat; the *Saint-Jean* then proceeded, fully loaded, to France in August 1668. On August 2, 1668, a second convoy consisting of the *Couronne* and the hooker *Saint-Denis* were sent to India to secure sorely needed supplies for Fort Dauphin. The third convoy, consisting of the *Marie* and newly arrived ships l'*Aigle d'Or* and la *Force* left in October 1668. The *Marie* transported the Director de Faye who died shortly after arriving in Surat. That fleet, after spending time in Persia, left Surat in January 1670, refreshed in Fort Dauphin in the spring of 1670 and proceeded to France on April 15, 1670. They brought back Governor Mondevergue who was later disgraced and imprisoned because of the disastrous situation of the settlement at Fort Dauphin.

In the meanwhile, the five hookers were used to sail back and forth to procure rice and oxen in Foulpointe, St. Marie and the Bay of

Antongil to supply the colony in Fort Dauphin. The petit *Saint-Jean* was sent in late 1667 on an expedition to explore the northeast part of Madagascar between the Bay of Antongil and the northernmost point. She was also to anchor south of the Bay of Vohemar to inquire about the fate of the three sailors of the wrecked *Taureau* who were left there. However, she had to interrupt her voyage about 100 miles north of the Bay of Antongil because a large part of the crew got ill. They were informed of the fate of three marooned sailors: one got murdered and the two others died from wretchedness. The petit *Saint-Jean* anchored in Foulpointe in late January 1668 before returning to Fort Dauphin. The coastal supply runs continued during 1668 but rice was scarce because locusts invaded the country just before harvest time. They had practically destroyed the crop. Fort Dauphin was hungry.

A fourth convoy, consisting of the *Couronne,* the frigate *Mazarine* and the hooker *Saint-Jean* left Fort Dauphin on August 12, 1669 bound for Surat and tasked to bring supplies to Fort Dauphin. The three ships rounded the south of Madagascar heading for the Mozambique Channel, but were separated by a storm a few days after their departure. The *Couronne* and the petit *Saint-Jean* managed to reunite and reached Surat on September 23. The *Mazarine* joined much later and was in such a sorry state that she had to be laid up. In late 1669, the petit *Saint Jean* sailed back to Fort Dauphin bringing supplies to the colony. On March 16, 1670, just after leaving Fort Dauphin for another supply run to India a furious wind pushed the petit *Saint-Jean* against the reef of the Banc de l'Etoile on Madagascar's southern coast. She immediately broke up and her cargo of 44 cannons, anchors, sail and rope was lost. Only one sailor of her 35-strong crew drowned in the wreckage. The survivors were rescued by barges sent from Fort Dauphin.

The *Saint-Luc* was kept in Fort Dauphin and carried out coastal trade to supply the colony. She foundered in a fierce storm at Itapere about 5 leagues northeast of Fort Dauphin on January 28, 1671.

LA DUNKERQUOISE (1674)[123]

In early 1673, the *Barbault* came back to France from Ceylon with the good news that Admiral Jacob Blanquet de la Haye had taken Trincomalee from the Dutch. Her commander, Captain de Beauregard was immediately named governor of the new possession in Ceylon.

He was ordered to return immediately with the same ship renamed *La Dunkerquoise* in order to occupy his new post, fortify it and make it flourish. His orders were to pass through Bourbon along the way to relieve Governor de la Hure who had caused so much grief and to reinforce the young French colony by disembarking a number of settlers, a cargo of tools and seed and sixteen young ladies chaperoned by Mademoiselle de la Ferrière, a nun.

By this time the 100 French settlers in Bourbon were in dire need of spouses. The ratio of men to women was about 3 to 1. Moreover most women were married Malagasy ladies. French ladies represented less than five percent of the male population. The settlers sent numerous petitions with the support of Governor Regnault asking for more ladies. The message was eventually heard in France. They decided to strengthen the colony in Bourbon by sending women. Who were these ladies bound for a new life on a faraway island? They were not really volunteers but the result of a raid undertaken by the East India Company at the Salpêtrière in Paris. This institution kept close guard on a disparate female population consisting of prostitutes, delinquents, orphans or young ladies given up by penniless families.

The *Dunkerquoise* left Rochefort in Brittany on May 29, 1673 for the 8 month trip to the Indian Ocean. Against his orders to sail directly to Bourbon, Captain de Beauregard decided to do a side-trip to Fort Dauphin to sell some of the large cargo of spirits taken on board for his own account. He anchored there on January 14, 1674 and found the remaining 127 French settlers in a destitute state. Nearly all of the settlers consisted of bandits and lazy and rebellious insurgents. He tried to convince them to move to Bourbon but the settlers refused to abandon the land for which they had given up so much. The sales of spirits was not very successful given the impoverished state of the settlers. Nonetheless, Captain de Beauregard procrastinated. The young ladies had gone ashore and the settlement was inhabited for the most part by single men so what needed to happen did happen: soon six of the sixteen ladies announced that they wished to remain there and get married.

Captain de Beauregard opposed this. His strict orders were to bring the ladies to Bourbon. Two weeks later he decided to set sail. This was the only way to stop this female hemorrhage that could cost him dearly. But the young ladies were in hiding; disappearing as by magic. In no instance could he leave without them. To make matters worse some of his men wanted to stay also. After remaining at anchorage for close to two months and with the weather worsening perceptibly, Captain de Beauregard announced on March 5, 1674 that he would leave with whoever wanted to follow him. When he was about ready to sail a fierce storm broke out. On board in total confusion some begged the Captain to cast off the moorings and move to the open sea as fast as possible. According to them this was the only way to avoid running aground. Others pressed for swinging out the boats and rowing back ashore notwithstanding the dangers involved.

This indecision lasted for more than 24 hours. The ship rocked violently and pulled all its weight on hawsers that threatened to snap with each motion. The tempest did not subside. One of the hawsers snapped and an anchor was lost. In these dramatic moments nobody understood why the Captain did not heed any of the recommendations offered to him and refused to take any initiative. Shortly thereafter, another hawser snapped, a second anchor was lost and the ship was tossed around in the storm even more. The hours went by and nobody predicted a favorable ending.

On March 7, 1674, the storm intensified even more. The third and last hawser snapped. The *Dunkerquoise* was now at the mercy of the pounding surf. In the ensuing chaos and to the dismay of everybody on board the first one to shamefully leave the ship was the Captain. The bravest crew members took over. They organized the evacuation of all the passengers as well as they could in these circumstances. After everybody disembarked the *Dunkerquoise* was carried by the current and thrown on the rocks where she broke up and immediately sank. A fitful and dramatic sight. On shore the crew did a roll call and blissfully found everybody was saved.

The Captain was disgraced. The whole community in Fort Dauphin heaped shame on him. The only ones who benefited from this sorry state of affairs were the young ladies. Now that it was impossible to bring them to Bourbon within a reasonable timeframe they let love run its course. Governor de la Bretèche consented to the marriages. If a ship were to pass by, they could be sent to Bourbon with their new husbands and strengthen the colony there hence fulfilling the King's wishes. During April and May six of the young ladies got married. However, these weddings created an ill-fated agitation with the surrounding Malagasy population: most of the settlers had mistresses, many had children from them and jealousies quickly surfaced. In June 1674, an epidemic broke out killing a number of the passengers, including three of the young ladies. There were only seven single ladies left and the jealousies grew to a boiling point. The young French ladies rejected overtures of the Malagasy men and the French settlers abandoned their Malagasy mistresses to elicit the favors of the young French ladies.

On August 9, 1674 a vessel of the East India Company, the 500-ton *Blanc Pignon* commanded by Captain Baron, anchored in the Bay of Fort Dauphin. His orders were to transport funds to India for the Crown. On the way back he was to assist the *Soleil d'Orient* that was experiencing trouble on the Mozambique coast. The layover was supposed to be short. However she offloaded 250 sailors deprived of women for many months. Their charms created havoc among the Malagasy community where tensions were already running high. As a result, the Malagasy decided to concoct a terrible vengeance. In a hurry to fulfill his orders and sensing that things were deteriorating rapidly, Captain Baron called all his sailors on board, weighed anchor and sailed on August 27.

As soon as the *Blanc Pignon* left the bay in the evening of August 27, the Malagasy attacked the settlers guilty of the sins of love. They were taken totally off-guard. It became a bloodbath. Seventy five settlers were killed, including two of the newly-wedded couples and two of the new grooms. The ones who could took refuge in the small fort and sent distress signals to the *Blanc Pignon* still wading offshore Fort Dauphin due to slack winds. Catching sight of the signals she turned around and slowly cast back in the bay.

Governor de la Bretèche knew that numerous settlers had fled to the neighboring woods at the time of the massacre. It was out of the question to leave without them. Captain Baron grudgingly agreed to wait for a few days. The next week was marked by search parties scouting the woods in search of survivors, punctuated by fights between the French and the Malagasy. In the evening of September 9, Governor de la Bretèche assembled the 63 survivors, set fire to the remaining installations at Fort Dauphin, pinned the cannons and embarked aboard the *Blanc Pignon*. Some French settlers had fled westwards and reassembled at St. Augustine Bay where they were picked up later by passing vessels.

Having lost a lot of time in Madagascar, Captain Baron decided to invert his itinerary and proceed to the Mozambique Channel first to lend assistance to the *Soleil d'Orient,* whose crew was decimated by yellow fever, before proceeding to Surat. It was slow sailing due to the very strong southern currents and the *Blanc Pignon* had to put in at Licongo for repairs. The *Blanc Pignon* finally reached Mozambique in May 1675 taking nine months instead of the usual two.

There, Captain Baron found only seven survivors of the 300-strong crew of the *Soleil d'Orient* but no ship.

Unbeknownst to Captain Baron, the Captain of the *Soleil d'Orient* had stuck a deal with the Portuguese governor in Mozambique to exchange some of the cargo for a fresh Portuguese crew and the *Soleil d'Orient* had sailed in July 1674 while the *Blanc Pignon* was nearing Madagascar. The *Blanc Pignon*'s crew and passengers received a warm welcome particularly since the Portuguese colony was also dreadfully short of young ladies. Charmed by the reception, two of the ladies announced their intention to remain there and marry Portuguese. To crown it all, the Portuguese Governor sequestered the six remaining single ladies! Only two of the original 16 ladies, both married in Fort Dauphin, left with the *Blanc Pignon* for Surat where they arrived in October 1675 together with their husbands and 20-25 other survivors from Fort Dauphin.

They all wanted to go to Bourbon as soon as possible but there were no ships plying that route. They had to wait until April 1676 to find a small hooker, the *Saint-Robert* commanded by Captain Auger, who was willing to take them. They finally reached Bourbon in late May 1676, three years after leaving France. One of the ladies was Nicole Coulon who had eight children with her husband Pierre Martin. The other was Françoise Châtelain who married Jacques Le Lièvre a settler from Fort Dauphin. Her husband was killed by marooned slaves two years later freeing her up for other suitors. Eventually she had three other husbands, Michel Esparon, Jacques Carré and Augustin Panon and a total of 10 children from them. This caused quite a bit of gossip. She tended to marry very soon after becoming a widow and one of her children was born two months after one of her weddings. Even so, these two ladies who suffered a terrible ordeal during their passage died with about 50 grandchildren between them. They thus belatedly fulfilled the wishes of Louis the XIV.

SOLEIL D'ORIENT (1681)[124]

The *Soleil d'Orient* was the first vessel built in the naval yards set up by Minister Colbert at Faouëdic in 1671. At 1,000 tons and armed with 60 cannons she was one of the largest vessels ever built for the French East India Company. Both the naval yard and the nearby city were named after her as people came to see the building of the *Orient*. Alas, she suffered an ill-fated history. She was launched at Port-Louis

Figure 64. Model of the *Soleil d'Orient* (Musée de la Compagnie des Indes, City of Lorient / Y. Boëlle).

on March 6, 1671 under the command of Captain Labeda but she lost her rigs and mast in a storm 100 leagues from the port. She put in at La Rochelle to be rerigged. Thence, the *Soleil d'Orient* sailed back to Lorient to wait for several months for the good season before proceeding. She finally sailed on March 12, 1672 under the command of

Captain de Boispéan bound for Surat and returned in August 1675 having experienced problems off Mozambique on the return voyage from India.[125] She was ready to sail again in 1677 but was prevented from leaving because of the Dutch blockade. After the Nijmegen treaty of 1678 reinstated the freedom of the seas she left for Surat on February 1, 1679 and for the next two years was used for the coastal trade between the different French factories in India.

In mid-1681, the *Soleil d'Orient* was ready to sail for the return trip to France with a rich cargo of spices, cotton and silk cloth, gold, silver and gems. Also on board was a solemn embassy from the King of Siam. In the mid-1670s the King of Siam, Naraï, had a vision of forming an alliance with France through King Louis XIV. Hoping to flatter and impress the French monarch, his Siamese counterpart prepared to send him a ship laden with priceless gifts. The Siamese ambassador, Phya Pipat Kosa and his deputies, Luang Sri Wisan and Khun Nakôn Wichai were designated in 1678 by King Naraï to travel to France. This royal decision was not to the liking of all. Indians of Ayutthaya, most likely the Brahman council, foretold a tragedy to Siamese who ventured afar on a foreign vessel. The King nonetheless kept to his decision and the embassy was ferried by the *Vautour* from Siam to Bantam (Java) in late 1680.

The *Vautour* was too small to undertake a long sea voyage while overloaded with the wealth of presents, the three ambassadors, the first ambassador's son, a delegation numbering twenty, a missionary and a young priest. On top of two young elephants, the treasures on board consisted of items considered valuable antiques even then. These included cabinets, chests, curious tables and boxes from Japan, gold and silver vases from China, a dinner service of gold, comprising over one thousand pieces that the Siamese king had received from the Emperor of Japan. There were letters engraved on gold leaf one and a half feet long and eight inches wide encased in golden chests encrusted with precious stones. Upon its arrival in Bantam, the embassy was installed in a house next to the French lodge whose walls were decorated with painted canvas and the floor covered with large carpets.

The lack of a proper vessel made the wait in Bantam last several months. The bad omen seemed came through even more as the third ambassador suffered some kind of paralysis. Eventually, the larger *Soleil d'Orient* turned up from Surat allowing the embassy to board.

The *Soleil d'Orient* departed from Bantam on September 6, 1681. Her precious consignment accompanied by the three Siamese ambassadors, wearing tall hats and smoking heavily. On October 1, 1681 she put in at Bourbon to undertake sorely needed repairs as she was very leaky. On November 1, she set sail once again never to be heard of or seen again. Louis XIV sent a ship from France to search for her to no avail. Some surmised that she was lost in a storm. In Bourbon, a priest imagined that the ship had caught fire remembering the incessant smoking of the ambassadors. Others believed that the ship was captured by the Dutch. The disappearance of this ship, on top of the numerous lives cost the Company 600,000 pounds (about $1.5 billion today) excluding the cost of the vessel.

In 1685, the Governor of the French East India Company in Bombay heard an interesting tale recounted by a French sailor. The sailor, Croisier, from Morlaix had boarded the *Coche* as a volunteer. In Surat he jumped ship and joined a British interloper (i.e. without license from the British Crown) the *Bristol* to return to Europe. The ship left Bombay in 1682 and stopped at Fort Dauphin for refreshments. Unable to round the Cape the *Bristol* put into the island of Anjouan (Comoros). While plugging a leak in her keel she was seized by the Admiralty vessel *Phenix* and brought to India. Taken by a storm she sank three leagues from Bombay bay. The crew was saved except for one man.

Croisier told the Governor that while in Fort Dauphin he made the acquaintance of a Malagasy servant called Jean who spoke fluent French as he had worked for the Governor in the days before the French settlement was forsaken. Jean told Croisier that the *Soleil d'Orient*, badly leaking, had put into Fort Dauphin in November 1681. He said the ship's occupants had befriended the locals and set about repairing their vessel. When the leak was plugged the *Soleil d'Orient* was ready to leave Madagascar. Surprised by a cyclone, she weighed anchor in a great hurry but broke up on the rocky coast somewhere northeast of Itapère with all hands lost.[126]

Without news from his first embassy, King Phra Naraï organized a second one in 1684. It was composed of Khun Pichaï Walit and Khun Pichit Maïtri. They arrived in France in October of the same year. They had an audience with King Louis XIV on November 27, 1684. The embassy of the chevalier de Chaumont brought these two ambassadors back to Siam.

Several marine archaeologists have searched for the remains of the *Soleil d'Orient* including Henri-Germain Delauze and Robert Sténuit in 1985 and Sverker Hallstrom in 1996. So far the wreck has not been found.

LE VAUTOUR[127] (1725)

The *Vautour* was a small 80-ton corvette mounted with 6 cannons and built in Lorient. She left on her maiden voyage for Bourbon on October 11, 1724 under the command of Captain Guillaume de la Butte-Frérot and 22 crew members. She was used as a supply ship and slaver between Bourbon, Mauritius and Madagascar. Her dramatic story encompasses one of the few successful slave uprisings at the time.

On July 5, 1725, the Company's Higher Council in Bourbon decided to send two small corvettes, the *Vautour* commanded by Captain La Butte, and the *Ressource* commanded by Captain Boulanger to Madagascar. Their instructions were to salt meat in Fort Dauphin and load animal fat, rice, woven pandanus, coffee bags and slaves. The *Ressource* was instructed to return to Bourbon immediately after being loaded up. The *Vautour* was allowed to secure slaves on Madagascar's west coast as Fort Dauphin was not suited for this kind of trade. After the two ships had left another small ship, the *Alcyon,* arrived in Bourbon from France earlier than expected. She was sent in early August to join the two other ships in Fort Dauphin. She was under the command of first mate de Selle as her Captain was ill. The instructions were for Captain La Butte to take command of the *Alcyon* and for first mate Le Marié to take command of the *Vautour.* Things went according to plan with the *Ressource* returning to Bourbon in September and the *Vautour* and the *Alcyon* securing slaves in the Morondava region on Madagascar's west coast.

On October 26, 1725 Captain La Butte and a few men joined by a few men of the *Alcyon* prepared for their imminent departure. They went ashore in their longboats to provision fresh water and spent the night ashore. Fourteen crew members remained aboard the *Vautour.* On October 27 shortly after 6 am, the 42 slaves aboard the *Vautour* mutinied. They were brought to the main deck every morning to get some fresh air and wash the deck. This day they took advantage of the master and the carpenter sneaking off for a drink. They seized the carpenter's axes, an adze and other metal objects that the cooper had neglectfully left in a barrel. They fought the crew members who were on deck killing the doctor, the carpenter, the cooper and a sailor. The first mate and acting Captain Le Marié was hit on the back of the head but escaped death by climbing the mizzenmast shroud and the yard. From there he alerted the crew of the *Alcyon* that was moored close by. The master climbed the foremast, two other sailors escaped by climbing the main mast and one sailor ducked in the steerage. The mutineers then cut off the moorings and let the ship drift.

Meanwhile, Captain La Butte watering ashore was warned by cannon shots from the *Alcyon* that something was amiss. Seeing that the *Alcyon* had put her flag at half-mast and that the *Vautour* was drifting he immediately dispatched the *Alcyon*'s doctor and a seaman in two dugouts. They approached the *Vautour*, but seeing armed slaves they went to the *Alcyon* for reinforcements, then returned close to the *Vautour*. In the meantime, the mutineers felt a slight jolt and anchored her. They pushed back four crewmembers who were trying to defend the main cabin and who retreated to the hold. Then they broke up the main cabin's walls and forced their way in seizing all the firearms and ripping open a large number of chests, barrels and bales of merchandise. Back on deck they started firing on the dugouts. Both capsized and one of the sailors was killed.

In the meantime, Captain La Butte had arrived with the two longboats. He picked up the surviving sailors from the overturned dugouts and went for reinforcements to the *Alcyon*. The mutineers saw the longboats come back and being well armed. They cut the hawser and shortly thereafter the *Vautour* beached. The mutineers fled ashore. Some were on two dugouts that had come to help, some swam, some were on hatches and others used boxes with arms and ammunition as a float. Captain La Butte and his men chased the fugitives with the longboats, killed a few but captured none.

The surviving crew tried to salvage the *Vautour*. The spare topmasts and yard were thrown overboard to lighten her. The sails were backed but winds that were blowing straight ashore impeded the operation. The crew of the *Alcyon* dropped one of her anchors and hawser some distance astern. Aboard the *Vautour* the crew tried to heave the ship off but she would not budge. By two in the afternoon, an intense surf battered the ship and she hove round broadside. With the stern already damaged, attempts to refloat the *Vautour* had to be abandoned.

In the ensuing days, Captain La Butte and his men cut the decks, joints and dead work on the starboard side to make sure that the wreck could be of no use to anyone. The merchandise that could be salvaged was stored ashore in a warehouse under the guard of two crewmembers. The *Alcyon* set sail for Bourbon on November 1 with the surviving crew members. She came back in August 1726 to pick up the merchandise and the two guards.

SAINT-PIERRE AND SALAMEC[128] (1746)

These wreckages are largely part of a naval campaign against Britain. Starting with the regency of Philippe d'Orléans (1715-1723) until the 1730s there was a tacit understanding and cooperation between the French and British merchant fleets on the sea route to India. This broke down with the advent of the Franco-British War because of the succession of the Austrian Empire followed by the Seven Years War. In 1741, Bertrand François Mahé, Count de la Bourdonnais, Governor of the Ile de France (Mauritius) and Bourbon from 1735 to 1740 commanded a royal fleet to fight the British dominance in India and to reinforce the settlements in Pondichery and Mahé.

After an uneasy truce war with Britain was declared in March 1744. De la Bourdonnais was again tasked with preparing a fleet to fight the British. He retained the *Duc de Bourbon* and the *Neptune*[129] that were calling at Mauritius and added the *Renommée* and the *Parfaite* that were used in the 1741 expedition and the locally built 320-ton frigate l'*Insulaire*. As these ships were ill-equipped in men and merchandise to take on the British, de la Bourdonnais used the first half of 1745 to train the men and arm the ships. On May 18, 1745 the French East India Company sent a fleet of five trading vessels armed as warships to Mauritius: the 1,200-ton *Achille,* the 700-ton *Saint-Louis,* the 800-ton *Phénix,* the 700-ton *Lys* and the 650-ton *Duc d'Orléans*. They were expected in the fall of 1745 but were delayed because of storms and arrived in late January 1746.

In the meanwhile, de la Bourdonnais took the necessary measures to supply and equip the fleet. Mauritius and Bourbon were unable to provide the needed supplies as they had suffered famine the year before. He requested naval equipment and provisions from India and secured more food in Madagascar. He sent the *Parfaite,* and two small vessels from the Indian fleet, the *Saint-Pierre* and the *Salamec,*[130] to Foulpointe on Madagascar's east coast to secure rice. On March 24, 1746 his fleet of merchant ships armed with a total of 310 cannons left Mauritius bound for Pondicherry.[131] Having only 65 days of food supplies on board, he decided to join the three ships securing supplies in Madagascar first. After a short stop in Bourbon to embark a number of loyal slaves the fleet arrived in bad weather in the morning of April 4 at Foulpointe.

On anchoring, de la Bourdonnais was informed by the crew of the *Parfaite* that the *Saint-Pierre* had wrecked on the coast and only the Captain and a few of the men had survived. In February 1730, the *Saint-Pierre* was sent from India to bring supplies to Bourbon. She was used for a couple of years on regular trips to Madagascar to secure food for Bourbon before being sent back to India. In early 1745, de la Bourdonnais ordered the *Saint-Pierre* and the *Salamec* to Mauritius as support ships for the fleet that he was assembling.

With a gale intensifying at 2 pm on April 4, de la Bourdonnais ordered all the ships to cast off. After nightfall the cyclone hit in full force and the furious sea impeded their progress towards St. Marie Island. At 10 pm, the *Lys* sent a distress signal: her foremast and bowsprit were broken off. The admiral ship *Duc de Bourbon* was also battered, losing her foremast, topmast, mizzenmast and bowsprit. At one point, they spotted the *Neptune*, which had two masts left. After raging all night the storm calmed down in the morning and the *Duc de Bourbon* and the *Lys* were able to reach the entrance of the Bay of Antongil. By 10 pm on the 5th they anchored at Marosy Island (Nosy Mangabe). The other ships joined them in the following days except for the *Salamec*.

The *Salamec* was lost in the raging storm in the early morning hours of April 5 at Mananara just south of the entrance of the Bay of Antongil. A couple of weeks later, the *Parfaite,* that had gone out to secure supplies, spotted the wreck and brought back her crew and passengers together with her masts, lifting gear and other accessories to Marosy Island on May 2.

La Bourdonnais spent two months in the Bay of Antongil to undertake the necessary repairs and to secure supplies. The *Parfaite* was sent back to Mauritius to bring the news of their misfortunes. The fleet sailed on June 1 and arrived in Pondicherry in early July after skirmishes with the English fleet at Negapatam. From there, de la Bourdonnais captured Madras on September 21, 1746.

CERF AND PHELYPEAUX[132] (1757)

The 230-ton corvette *Cerf* was built in Lorient and mounting 8 cannons was launched on September 11, 1751. She left for Senegal on her maiden voyage on May 21, 1752 and returned in November of that year. On June 14, 1755, she left Lorient bound for Mauritius with a crew of 51 under the command of Captain François Barbotin arriving on February 10, 1756. Next she sailed to Bourbon. On July 16, 1756, the *Cerf* with Captain Nicolas Morphey and the *Saint-Benoît* left Bourbon with orders from King Louis XV to explore Praslin island in the Seychelles.

However, violent currents pushed both ships 25 miles to the south forcing them to anchor in the Bay of Mahe on September 6. They explored the surrounding forest and only found crocodiles and giant turtles while being surrounded by thousands of multicolored birds. The island was uninhabited. On November 1, 1756 a carved stone bearing the royal coat of arms was placed at the entrance of the bay and Captain Morphey officially took possession for France of the Archipelago of the Seychelles named in honor of Jean Moreau de Séchelles. He was the French Minister of Finance at the time and had financed the expedition. The two vessels left Mahe around mid-November, visited the Farquhar and Cosmoledo islands on the way in, and arrived to a warm welcome in Bourbon on December 10, 1756.

On December 29, 1756, the *Cerf* was sent to Madagascar under the command of Captain Gérard to recover a cargo of salt, which was needed in Fort Dauphin for salting meat and which had been left behind on St. Marie Island by the vessel *Auguste.* While in St. Marie the *Cerf* was taken in a cyclone. The crew cut the hawsers and moved to the open sea, but the violence of the storm threw her on the rocks at Foulpointe on January 17, 1757. Fourteen crew members died in the wreckage and 38 made it alive.

Less than a month later on February 13, 1757, another vessel, the 650-ton *Phelypeaux,* built in Bayonne in 1751 and mounting 18 cannons also foundered in St. Marie. The *Phelypeaux* was launched on July 24, 1752 and sailed from Rochefort to Lorient. On April 8, 1753 she left Lorient bound for Pondichery under the command of Captain Nicolas Claëssens returning to Lorient at the end of the following year. She left Lorient for a second voyage on March 19, 1755 bound for Mauritius under Captain Jean-Joseph de Sanguinet with 134 crew members.

In mid-1756, the *Phelypeaux* was sent from Bourbon to Madagascar's west coast to procure slaves, whose number had dwindled in Bourbon and Mauritius because of an outbreak of smallpox. Captain de Sanguinet acquired 337 heads. On the return trip, for some unexplained reason, he missed both Bourbon and Mauritius and then sailed to St. Marie Island to resupply. While anchored in the harbor of

St. Marie, the *Phelypeaux* was taken in a violent storm and foundered. There was no loss of lives.

In April 1757, the *Rubis* was sent from Bourbon to St. Marie Island to pick up the crew, passengers and slaves of the two vessels and what could be saved from the cargo.[133] Meanwhile, 100 of the salves had died in St. Marie, and another 43 died shortly after their arrival in Mauritius.

GLOIRE[134] (1761)

The 311-ton frigate *Gloire* mounting 24 cannons, was built in Lorient and launched on February 1, 1753. She left for her maiden voyage to Pondicherry on April 8, 1753 returning in March 1755. She underwent a major careening in Lorient between April and August 1755. The *Gloire* sailed for a second voyage under Captain Michel Hay bound for Surat on August 8, 1755. She carried letters with the bad news of the deteriorating relations between France and England and instructions for the factories to start preparing for a possible conflict.[135] She was laid up in Mauritius on the return of this voyage under the command of Captain Jacques des Essarts on December 31, 1756. The *Gloire* was again commissioned in August 1757 and used for coastal navigation in the southwestern Indian Ocean. She undertook a run to the Bay of Antongil in the summer of 1761 under the command of Captain Nevé anchoring in Venaguebe Cove (North of the Bay of Antongil) to trade rice.

In late July, the *Gloire* was loaded and ready to sail. But seaward winds being rare in Venaguebe she had to wait a few days before being able to cast off. On August 4, 1761 the winds were favorable and Captain Nevé ordered to weigh the anchors. Just after crossing the channel where they had anchored the winds died down. The ship became unsteerable and floated dangerously close to the rocks. Captain Nevé anchored and had the cables extended to move away from the danger zone. Suddenly, the winds freshened and became violent. The *Gloire* started dragging on her anchors. Captain Nevé ordered the cables to be cut and to fill the sails in order to reach the bay. However, most of the crew was drunk and reaction times were too slow. The wind threw the *Gloire* against the rocks and she ripped open. All of the cargo was lost. The crew was saved and went back to Mauritius on various ships.

FORTUNE (1775), SIRÈNE(1776), COUREUR AND INDIGENT[136] (1777)

The next French attempt to conquer Madagascar took place about a century after the August 1674 debacle at Fort Dauphin. Save for a short-lived attempt by the Count de Maudave in Fort Dauphin in 1768 it came about through the private initiative of the Polish Count Benyowsky in the name of the French King Louis XV. Benyowsky was a Hungarian nobleman of Slovak origin. He began his career as an officer in the Seven Years' War then joined the Polish Confederation in its fight for independence from the Russian rule. After being captured by Russians in 1770 he was sent into exile to east Siberia (Kamchatka). He managed to escape and capture the fort of the governor and the heart of his daughter. He commandeered a Russian battleship and set out for a discovery trip through the Northern Pacific along the Aleutians, Alaska, Japan, Formosa (Taiwan), Canton arriving in Macao in 1771 (well before James Cook and J. F. La Perouse). The next year, he embarked on the French vessel *Dauphin* bound for Lorient. On the way he visited Mauritius and Madagascar then still independent and ruled by countless native chieftains.

On arriving in France Benyowsky suggested to King Louis XV that he should establish a French colony in Madagascar. The King entrusted him with an expedition to Madagascar and appointed him as Governor of the island. Benyowsky left France in April 1773, spent a few months in Mauritius late that year and arrived in Madagascar in February 1774 with a corps of volunteers. He established a colony at Maroantsetra (Antongil Bay) called Louisbourg with a hospital and quarantine on Nosy Mangabe. Besides building the French presence and geographically exploring the island, he unified tribes. In 1776 local kings elected him as their Ampanjakabe (Emperor). Among other things he introduced Latin script for the Madagascar language.

In the fall of 1774 Louis XV sent the navy flute *La Fortune* under the command of Captain Morel with men and supplies to strengthen Benyowsky's settlement. Reaching Madagascar in the spring of 1775 Captain Morel put in at Fort Dauphin. Having entered the bay with a very light wind he threw one anchor, then decided to have lunch with his crew neglecting to moor by the head and to sound the depth. Before lunch was over it became apparent that the cable had ripped, probably due to friction on a coral head. The ship was floating towards shore. Some confusion ensued. It took a while before a second anchor was dropped. The *Fortune* was then riding the wave while the winds had freshened. They swung out the cutter but before a small anchor could be affixed the *Fortune* was thrown ashore and immediately broke up.

At the beginning of his operation Benyowsky possessed only one vessel deputed for the service at Madagascar. It was the snow *Postillon* commanded by Captain de Saunier. She was used early on to sail back and forth between Antongil and Foulpointe to help set up an establishment there. Benyowsky's directives stipulated that privately-owned vessels were banned from trading in Madagascar. While supplying Foulpointe, the *Postillon* patrolled the coast. In August 1774 she seized the private vessels *Flore* and *Coureur* and escorted them to Louisbourg. Benyowsky bought the *Coureur* and its cargo of slaves in the name of the King. He needed that vessel to trade on Madagascar's east coast, even more so since on September 25, the *Postillon* was sent back to France to plead the case of the establishment and request more men and supplies. In late September, the *Coureur* under the command of Captain Desmousseaux was sent to Mauritius. Upon her return in mid-December, an aide to Benyowsky, the geographer Garreau de Boispréaux, embarked on her to undertake a study of Madagascar's southern coastline and of Fort Dauphin. He came back to Louisbourg in mid-March 1775.

After reaching an agreement with the owner, Benyowsky let the *Flore* go with a cargo of rice. However, he got wind that she had surreptitiously and against orders loaded a supply of slaves and was bound for Cape Town. On April 1, 1775 Benyowsky embarked on *Coureur* to chase the *Flore*. He did not need to go very far. The slaves had rebelled, managed to escape to the coast and the *Flore* was back in Foulpointe on April 20. Benyowsky impounded the vessel, purchased her in the name of the King and renamed her *Maurice Auguste*. He sent her immediately to Mauritius with his family to ask reinforcements for the imminent war against the Sakalava tribe. In late 1775, *Coureur* was sent to Mozambique to secure slaves. She returned to Louisbourg on January 11, 1776 after a call at Mauritius confirming the vanishing support of the Governor for Benyowsky's venture in Madagascar.

To add insult to injury, the corvette *Sirène,* a new ship that was to replace the *Postillon* to serve the needs of the establishment and had left Lorient in August 1775 with 60 men and merchandise, wrecked in the south of Madagascar close to Fort Dauphin, on January 28, 1776. Captain de Saunier had committed a critical navigation error. He failed to slot in the significant currents prevailing in these waters and he believed he was sailing about 50 leagues from Madagascar's east coast. He threw his ship on the rocks strewing the southern end of the island. The bulk of the crew managed to reach shore except for 10 men who drowned. The survivors headed towards Fort Dauphin. The only weaponry they possessed consisted of six guns, three swords and two hunting knives. A few hours out, they were attacked by 500—600 warriors. Unable to put up a defense, they were robbed of everything they had, separated and ushered to various villages. They were eventually freed and reassembled in Fort Dauphin. There illnesses took a toll. Most died including Captain de Saunier. A few survivors were picked up by a private merchant ship in early August and brought to Mauritius.

Benyowsky kept pleading for more resources. Though he was promoted to French General Paris ignored his requirements. Brokenhearted, Benyowsky left Madagascar on December 14, 1776. Accused of mismanagement and disgraced in France, Benyowsky returned to Central Europe. In 1779 he went to America to help the revolution. With Benjamin Franklin's assistance he founded an American-British company for business with Madagascar. In 1784, Benyowsky left Baltimore on board of the *Intrepid* provided by Baltimore businessmen Messonier and Zollickofer. Back in Madagascar he challenged the European powers in the name of his empire. This episode ended tragically on May 24, 1786 when a French detachment sent from Mauritius under the command of the knight de Tromelin assaulted him. Benyowsky perished as Malagasy monarch in that battle.

After Benyowsky's departure, the remnants of the settlement in Madagascar were run by some of his deputies and headed by the Chevalier de Sanglier. Things deteriorated quickly as many settlers died from fevers and other diseases. The *Coureur* continued to be used for coastal trade. She foundered in a storm six leagues from Fort Dauphin in mid-1777. Returning from an inspection tour in Foulpointe in August 1777, de Sanglier encountered close to Fenerive Captain Ravisseau, the officers and crew of the private brick *l'Indigent*. He was told that the ship had wrecked in the north of Baldriche Point close to Fenerive. As the *Indigent* started to sink due to a massive water intake Captain Ravisseau decided to beach her with the ensuing loss of all of the cargo. Sanglier sent the Captain, officers and crew to Mauritius on two vessels berthed at the time in Antongil, the *Triton* and the *Mimie*.

COMTE DE MAUREPAS[137] (1777)

This 900-ton vessel mounting 36 cannons, was built in Amsterdam and originally named *Saint Jean-Baptiste*. She was purchased in June 1765 by the French East India Company and on that occasion renamed *Comte de Maurepas*. In April 1777, the shippers Bernier and Gourlande chartered her for a private run to India. On May 10, 1777 she sailed from Lorient bound for Bengal under the command of Captain Jean Muterse de Guérande with 112 crew and passengers on board. She was carrying 40,000 Pieces of Eight (about one ton of silver coins) for general purchases in India together with valuable personal effects of the VIPs on board. After an uneventful passage, they reached Madagascar's southeastern coast in July and were heading for the Bay of St. Augustine, where they intended to seek refreshments.

In late afternoon of July 19, the *Comte de Maurepas* passed Cape St. Marie (Madagascar's southernmost point) and Captain Muterse dutifully recorded this in his log. He then passed the wheel to one of the officers with specific instructions to steer her northwest ¼ north on the compass and descended to the main room to have supper. The officer either misunderstood the order, or was negligent and steered her on a north ¼ northwest course. This brought him dangerously close to the coast. In heavy seas, the officer suddenly saw an outcrop and sounded the alarm. Captain Muterse rushed back on deck, immediately sensed the danger, took the helm, and tried to steer the *Comte de Maurepas* to starboard, but too late. She violently crashed against the reef of Star Bank (Banc de l'Etoile) and burst open. In less than five minutes the hold was full of water.

The Captain, officers, VIP passengers, and some of the men, 69 in total, boarded the ships' boat and set out for the Bay of St. Augustine. The 40-odd remaining on board built rafts and let themselves drift to the coast close to Star Bank, never to be heard of again. Lack of fresh water prevented the boat to sail all the way to the Bay of St. Augustine. They landed on the coast 40 leagues north of Star Bank, then walked overland to St. Augustine, arriving there on July 25. Thirty eight of the 69 survivors died from tropical diseases in the following months. The doctor on board and five other crew and passengers left the Bay of St. Augustine on August 9 and walked overland to Fort Dauphin. They arrived in early September. They were picked up by the ship l'*Inconstant* bound for Mauritius, and from there sailed back to France on the ship l'*Archangele*.

On January 4, 1778, the Captain and the 25 survivors walked from St. Augustine to the Bay of Toliary, 10 leagues to the north. There, all of them were picked up by a Dutch ship, the *Zagr Ruste,* on its way from Zanzibar to the Cape, except for one man who was terminally ill and left behind. On February 8, 1778 the Captain and four men boarded the *Sainte Anne* that refreshed at the Cape on her way from Mauritius back to France. Back home, Captain Muterse wrote a full report on May 8, 1778.

REFERENCES

[104] Sources used for this historical introduction include: Dufresne de Francheville, "Histoire de la Compagnie des Indes," 1740; de la Roncière, C., "Histoire de la Marine française," 1909; Grandidier et al. op. cit., t. VII; Van der Cruysse, D., " 'O navigants, o povres mathelotz' les échecs maritimes du XVIe siècle," in Le Noble désir de courir le monde. Voyager en Asie au XVIIe siècle, Paris: Fayard, 2002, pp. 15-20; French East India Company, Wikipedia at http://fr.wikipedia.org/wiki/Compagnie_fran%C3%A7aise_des_Indes_orientales; and Histoire de la Compagnie des Indes at http://enguerrand.gourong.free.fr/oceanindien/p010ceanindien.htm.

[105] Recorded in Correa, op. cit., t. II, pp. 241 and 385, De Barros, op. cit., Dec. IV, liv. iii, ch. ii, p. 261 and liv. v, ch. vi, p. 583, and Grandidier et al., op. cit., t. I, p. 59; and reported by Le Lan, J.-Y., "Le commerce français avec l'Asie avant la Compagnie des Indes Orientales de Colbert," Histoire Généalogie, 2003 at http://www.histoire-genealogie.com/spip.php?article158; and Le Bris, M., "D'or, de rêves et de sang, L'épopée de la flibuste (1494-1588)," Paris: Hachette, 2001, pp. 37-38.

[106] Together with three survivors of the *Conceição* and one from the *São Sebastião* (see above).

[107] Related in Froidevaux, H., "Les Préludes de l'Intervention française à Madagascar au XVIIe siècle, Navigateurs, géographes et commerçants français de 1504 à 1640," Revue des Questions Historiques, 1909, pp 9-11; Le Lan, J-Y, op. cit.; and Le Bris, op. cit., pp. 39-40. The journal of this expedition has been reproduced in Magry, P., "Journal d'une Navigation des Dieppois dans les Mers Orientales sous François 1er (1529-1530)," Bulletin de la Société Normande de Géographie, May-June and July 1883, pp. 168-248.

[108] Martin de Vitré, F., "Premier Voyage fait aux Indes Orientales par les Français, en 1602," pp. 11-12; and Froideveaux, H., op. cit., pp. 27-28.

[109] He wrote two books about Madagascar's history, mainly the "Histoire de la Grande Isle Madagascar," Paris: Alexandre Lesselin, 1658, with a second edition in Troyes: Nicolas Oudot, 1661. Claude Allibert published an annotated version of the 1661 edition, Paris: Inalco-Karthala, 1995.

[110] Caron, a Flemish Catholic by birth, joined the VOC as a young man, converted to Calvinism, and eventually became the Governor of Batavia, where he acquired an invaluable knowledge of China and Japan. After retiring from the VOC and settling in The Hague, he was hired by Colbert to head the new Compagnie.

[111] Sources: Kaeppelin, "Les Escales françaises sur la Route de l'Inde, 1638-1731"; Flacourt, op. cit., 1661 ed., Avant-Propos, p. 15-16; Grandidier et al., op. cit., t. II, p. 466-469 and 506-508, and t. VII, pp. 29-129; Leupe, "Verhandelingen en Berigten bettrekkelijk het zeewezen en de zeevaartkunde; verzameld en uitgegeven door Jacob Swart," ch. XXII, "De vestiging der Hollanders op Mauritius in 1638," Amsterdam: Hulst van Keulen, new series, 1854, pp. 265-281.

[112] The first Dutch settlers had arrived on March 7, 1638 and were building a fort in the southeast port.

[113] Lohatanjona Itaperina, north of Fort Dauphin, off the town of Evatra.

[114] Including François Cauche, who wrote a report of his travails, collected by Morisot, "Relation véritables et curieuses de l'Ile de Madagascar; Relation du Voyage que François Cauche de Rouen a fait en l'Ile de Madagascar," Paris: Augustin Courbé, 1651, most of which is believed to be a work of fiction. For a contrary opinion, see Lieutenant Gaubert, "François Cauche," Revue de Madagascar, 1903, April, pp. 289-305, and May, pp. 385-403, and Allibert, op. cit., p. 17.

[115] Sources: Flacourt, op. cit., pp. 193-196 and 1661 edition, p. 205-206.

[116] Sources: Log book of the *Duchesse*, Archives Nationales, Paris, Colonies, C/5A1/No. 2; Martin, Fr., "Extraits des Mémoires sur l'Etablissement des Colonies Françoises aux Indes Orientales, 1665-1668," Manuscript, Bibiliothèque Grandidier, Antananarivo, No. 2987, pp. 48-57; Centre des Archives d'Outre Mer, Aix en Provence, Correspondence générale, Madagascar, Carton 1, piece 2; Flacourt, op. cit.; Grandidier et al., op. cit., t. III, p. 282-284; Capitaine de Villars, "Madagascar 1638-1894. Établissement des Français dans l'Ile," Paris: L. Fournier, 1912, pp. 47-48.

[117] Sources: Souchu de Rennefort, "Histoire des Indes Orientales," Leide: Frederik Harring, 1688; Martin, Fr., op. cit.; de Villars, op. cit., pp. 59-70; Enis Rockel, "De l'île sans nom à l'île Bourbon ou, Bourbon des origines jusqu'à 1700," at http://amis.univ-reunion.fr/Conference/Complement/166_bourbon/index_bourbon.html.

[118] Formerly called l'*Aigle Noir,* she was sent by Fouquet in 1661 for his own interest, under the command of the Dutch Captain Hugo, officially to scour the Red Sea, but secretly to seize Madagascar. After being damaged in a storm off the Cape of Good Hope, she arrived in Madagascar, but Captain Hugo and his troops were prevented from taking Fort Dauphin by the defenses put up by Governor Champmargou. The *Aigle Noir* then sailed to the Red Sea, did some acts of piracy, returned to Madagascar, and sailed back to France with a cargo of hides, ebony wood, indigo, and spices. To whitewash this expedition, the

new Company renamed her *Saint-Paul* (see Souchu de Rennefort, op. cit., pp. 62-65).

[119] For an account of that event, see: Martin, Fr., op. cit., pp. 14-16; and Desperthes, J.L., "Histoire des Naufrages ou recueil des relations les plus intéressantes des naufrages arrivés depuis le XVème siècle jusqu'à nos jours. Nouvelle éd., refondue, corrigée et augmentée, by J.B.B. Eyriés," Paris: Ledoux et Touré, 1815, vol. 2, pp. 37-48.

[120] For more information on this wreckage, see http://www.treasurelore.com/florida/treasure_ships.htm. In May 2008, Odyssey Marine Exploration, a Florida-based deep-ocean shipwreck exploration company, filed a claim at the District Court in Tampa to search the vessel. For more details, see Milmo, C., "Pirates of the Channel Islands: A £200m treasure hunt," The Independent, June 12, 2008 at http://www.independent.co.uk/news/uk/home-news/pirates-of-the-channel-islands-a-163200m-treasure-hunt-845054.html.

[121] Sources: Journal held by Mondevergue, Archives Nationales, Paris, Marine, 4JJ/75; Souchu de Rennefort, op. cit.; Martin, Fr., op. cit.; de Villars, op. cit., pp. 71-73; Ruelle, "Voyage à Madagascar et aux Indes orientales," s.d., manuscript, Bibliothèque Grandidier, Antananarivo, No. 2723; Enis Rockel, op. cit.

[122] Sometimes called *Saint-Jean-Baptiste*.

[123] Sources: Souchu de Rennefort, op. cit., pp. 549-552; de Villars, op. cit., pp. 80-93; Ruelle, op. cit., pp. 50-53; Daniel Vaxelaire, Service Multimédia, March 4, 2005, at http://reunion.rfo.fr/article124.html; Enis Rockel, op. cit.

[124] Sources: Archives Nationales, Paris, Colonies C1 22, folios 4 through 38; Lanier, L., "Etude historique sur les relations de la France et du Royaume de Siam de 1662 à 1703," Versailles : E. Aubert, 1883, pp. 19-31 ; Launay, A., "Histoire de la Mission de Siam 1662-1811, Documents Historiques," Paris, 1920, t. I, pp. 112 and 138; Hutchinson, E.W., "Aventuriers au Siam au XVIIe siècle," [translated by H. Berland], Bulletin de la Société des études indo-chinoises, Vol. XXII, Saigon, 1947, pp. 65-66 ; Histoire de la Compagnie des Indes at http://enguerrand.gourong.free.fr/oceanindien/p01 0ceanindien.htm; Sverker Hallstrom, Shipwreck Explorer-Soleil d'Orient at http://www.shipwreckexplorer.com/hallstrom/soleil/soleil.htm; Enis Rockel, "Vieux gréements—Le Soleil d'Orient ," at http://www.locmiquelic.org/Bateaux/index.html.

[125] The *Blanc Pignon,* that had picked up the survivors of the Fort Dauphin massacre (see above), was bringing a replacement crew as most of the original crew of the *Soleil d'Orient* was stricken by yellow fever during the layover in Mozambique.

[126] Jean and Croisier's tale included, apart from several crew members' names, a reference to the presence of the three Siamese ambassadors, wearing their tall hats and smoking heavily, information that would have been impossible to know without the ship's actually having anchored at the port.

[127] Source: Compagnie des Indes Orientales, "Dossier relatif à la révolte a Madagascar, à bord du *Vautour,* des esclaves déjà traités par ce navire," 1725, 12

fol., Archives Départementales de le Réunion, C°1389; s.n., "Quelques documents concernant la perte du négrier *Vautour* à Madagascar, en 1725," Recueil Trimestriel de Documents et Travaux Inédits pour servir à l'histoire des Mascareignes françaises, Janvier-Mars 1937, No. 4, pp. 347-372.

[128] Sources: Letter written on May 5, 1746 by Mr. de la Bourdonnais to Mr. de Saint-Martin, Deputy Governor of Mauritius, reproduced in Grant, Ch., "History of Mauritius," London: G. and W. Nicol, 1801, pp. 241-245, and Grandidier et al., op. cit., t. V, pp. 242-247; Compagnie des Indes, "Vaisseaux avec lesquels M. de la Bourdonnaye a passé aux Indes," May 8, 1747, Centre des Archives d'Outre Mer, Aix en Provence, FM, C2-275, folio 101; Lougnon, A., "Vaisseaux et traites aux îles depuis 1741 jusqu'à 1746," Recueil Trimestriel de Documents et Travaux Inédits pour servir à l'Histoire des Mascareignes Françaises, Tome No. 1, Tananarive, April-June 1940, pp. 11-33; Sulliman, "Bourbon au secours des Indes," Clicanoo, le journal de l'île de la Réunion, March 13, 2005, at http://www.clicanoo.com/article.php3?id_article=99074.

[129] The 800 ton *Duc de Bourbon* had done three roundtrips from Lorient to India between 1735 and 1742. On January 1, 1744, under the command of Captain de Plaisance, she sailed from Lorient bound for Pondichery. The 700 ton *Neptune* had done two roundtrips from Lorient to China between 1740 and 1743. She left again in December 1743, bound for Bengal, under the command of Captain de la Porte-Barré. While calling at Mauritius in late 1744 on their return journeys, they were conscripted in de la Bourdonnais' fleet.

[130] Sometimes also called *Neptune*.

[131] The *Neptune* left Mauritius a couple of days ahead of the rest of the fleet to bring de la Bourdonnais' instructions dated March 22, 1746 to Bourbon (Lettre de M. Mahé de la Bourdonnais par le *Neptune* au Conseil Supérieur de Bourbon, 1 fol., Archives Départementales de la Réunion, C°386).

[132] Sources : Compagnie des Indes, "Etat Général des Vaisseaux de la Compagnie des Indes, tant à la Mer et en Europe, qu'en construction, avec quelques notes sur ce qu'ils sont devenus, jusqu'au 12 février 1758," Centre des Archives d'Outre Mer, Aix en Provence, FM, C2-275, folios 104-107 ; French Naval Archives (Service Historique de la Défense—Centre des archives de l'armement) ; Service Historique de la Défense, Archives de la Marine, Lorient, file 1 P 193; Eschapasse, B., "Les Français chassés du Paradis," Historia, at http://www.historia.presse.fr/data/mag/722/72202601.html.

[133] Compagnie des Indes, Lettre du Conseil Supérieur à celui de Bourbon, "Naufrage à St. Marie de Madagascar du *Cerf* et du *Phélipeaux* : expédition du *Rubis* sur les lieux," April 12, 1757, 2 fol., Archives Départementales de la Réunion, C°470.

[134] Sources: French Naval Archives (Service historique de la défense—Centre des archives de l'armement); Letter by Mr. Vally, Quartermaster of the *Gloire*, dated September 6, 1761, Archives Nationales, Paris, Colonies, C/5A2/28.

[135] Compagnie des Indes, Lettre du syndic et directeurs de la Compagnie des Indes au Conseil de l'Isle de Bourbon, August 8, 1755, 4 fol., Archives Région-

ales de la Réunion, C°162; ibid. au Conseil supérieur de Pondichéry, 2 fol., loc. cit, C°2916.

[136] Sources: Benyowsky, M.A., "Memoirs and Travels of Mauritius Augustus, Count of Benyowsky, Magnate of the Kingdoms of Hungary and Poland, one of the Chiefs of the Confederation of Poland, Consisting of his military operations in Poland, his exile into Kamchatka, his escape and voyage from that peninsula through the northern Pacific Ocean, touching at Japan and Formosa, to Canton in China, with an account of the French settlement he was appointed to form upon the island of Madagascar," Translated from the original manuscript by William Nicholson, London: G.G.J. and J. Robinson, 1790; and "Voyages et Mémoires de Maurice Auguste, comte de Benyowsky, magnat des Royaumes de Hongrie et de Pologne, contenant ses opérations militaires en Pologne, son exil au Kamtchaka, son évasion et son voyage à travers l'Océan Pacifique, au Japon, à Formose, à Canton en Chine, et les détails de l'établissement qu'il fut chargé par le ministère français d'organiser à Madagascar," Paris : F. Buisson, 1791; de Villars, op. cit., pp. 117-143 ; Vacher, P., "Contribution à l'histoire de l'établissement français à Madagascar par le Baron de Benyowszky (1772-1776), D'après de nouvelles sources manuscrites," Collection 'Clio en Afrique,' N° 19, été 2006, Editions du Centre d'Etude des Mondes Africains, MMSH, Aix en Provence, at http://www.mmsh.univ-aix.fr/iea/Clio/VACHER.pdf; Word of Wordland, "Count Matus Moric Benovsky," at http://www.slovakopedia.com/m/moric-benovsky.htm; Archives Nationales, Paris, Colonies, C/4/40/173-174, C/5A5/75 and C/5A8/26, 82, and 91.

[137] Sources: Jean-Yves Le Lan, "Le Naufrage du Compte de Maurepas," Histoire-Généalogie, at http://www.histoire-genealogie.com/article.php3?id_article=168, November 1, 2004; Service Historique de la Défense, Archives de la Marine, Lorient, file 2 P 15-I-7; Archives Nationales, Paris, Colonies, C/5A8/129.

5
Pirates

Arab, Persian and Indian pirates have roamed the Indian Ocean since ancient times. In these days eastern Madagascar was a secure heaven because it was comparatively little visited by other ocean-going shipping and was exploited as a safe place for pirates to return to, store up riches and repair ships. To quote John Mack: "Early sources talk of an attack on the island of Kanbalû which is likely to be one of the islands of the Comoros around the year AD945. The perpetrators of the attack from the sea were a large force of the so-called WaqWaq who are also said to have ravaged the eastern Africa coast as well. Modern opinion associates the WaqWaq with Indonesia and sees Madagascar as their likely base."[138] Other sources record privateers to have sailed in the vicinity of the Comoros and Madagascar around the 13th century.[139]

Beyond these early and fleeting stays, Madagascar became a stronghold of Western piracy from around 1680 to 1725. The origin of Western piracy is closely linked to the discovery of the New World by the Spanish in the late 15th century. In the 16th century, the English, Dutch and French started competing for the riches that the Spanish were taking from the Amerindians while vying for territorial gains in the Western Hemisphere. In this quest, they did not hesitate to provide special government licenses to privateers to support the national fleets in fighting the Spanish. Privateers held a letter of marque from their government authorizing them to seize enemy nations' ships and their cargoes. To quote Jean Hood: "Privateers were useful in that they were privately financed, bore all their own risks and thus assisted their country in wartime at no cost to the government. Prize ships were taken into service or sold along with their cargoes and the proceeds shared among those with a stake in the enterprise."[140] Privateers were thus seen as law-abiding sailors by their home country but were viewed as pirates by rival nations. Moreover, privateers often blurred the line between the legal and unlawful by launching unauthorized attacks in addition to the sanctioned ones. In doing so, privateers turned buccaneers or pirates, preying indiscriminately and without legitimacy.

With technological progress, maritime trade expanded significantly in the 17th century and became the foundation of an international commercial network. It also produced a previously nonexistent line of work for sailors. Life was typically hard and pay was meager for these sailors. Many were conscripted in merchant or naval fleets. Besides hard work, inadequate food and illnesses the punishments inflicted by the officers included being put in irons, flagellation and being dragged under the keel from one to the other side of the ship. This often had fatal consequences. Being conscripted on a ship was considered harder than a prison sentence. Conversely, the pirates had created their own world where they made their own choices unencumbered—a world of solidarity and brotherhood where the risks and rewards of life at sea were shared and where decisions were made in a collective fashion without toiling to make some rich merchant even richer. Not surprisingly, defections of merchant or naval sailors to join the ranks of pirates became commonplace.

With the consolidation of the colonial empires in the Western Hemisphere and the set up of steady and orderly trade patterns attitudes toward piracy started changing. Gone were the days of a fine line between legitimate trade and piracy and state-sponsored adventurers such as Sir Francis Drake. Pirates became a dejected bunch of mutineers, slaves on the run and multi-ethnic rebels. In reaction to this official rebuff and tracking down by imperial navies, pirates became more antagonistic and declared war on all states and their laws.

In the second half of the 17th century the Caribbean became a melting-pot of rebellious poor immigrants from the world over. Deported prisoners formed a relatively egalitarian dropout culture[141] and a hotbed for buccaneers. These classless principles were transmitted aboard pirate ships.[142] Leadership was assured by a Captain charged with steering the ship and commanding during battles; a first officer responsible for the discipline of the crew, supplies and the allotment of the booty. Pirate Captains were elected and could be dismissed at any time for abuse of authority. Captains lead but other decisions were made democratically among the crew which was not always very efficient. Each pirate ship had its written rules. They were adopted and signed by everyone on the crew who swore to abide by them by placing two fingers on the Bible. Catches were divided equally among the crew and solidarity, notably for the wounded, was taken very seriously. Pirates formed a commonwealth and developed their own language. They kept in touch, did not attack each other and often joined forces to undertake joint raids. They put in at "free ports" where they could sell their merchandise to black marketers, socialize and organize wild parties after capturing a rich booty.

This egalitarianism also covered the racial sphere and was accompanied by sexual tolerance. Whereas a few pirate ships are known to have participated in the slave trade, a large number of pirates were former slaves[143] and a number of black pirates became first mates and Captains. Some men joined the pirates' ranks to flee the profoundly anti-homosexual mores of that time as homosexual relations were usually tolerated aboard pirate ships.[144]

Toward the end of the 17th century piracy in the Caribbean was on the decline. Buccaneers were being driven out of their favorite haunts and hangouts—Tortuga was beginning to settle down under French rule and Port Royal (Jamaica) was still reeling under the shock of the Great Quake of 1692. The Spanish treasure fleets were less frequent and better armed and risks increased through better naval patrols. Stories began to reach the Americas concerning rich Moorish vessels waiting to be taken by anyone enterprising enough to make the journey across the Atlantic and into the Indian Ocean. Before long, many American and Caribbean pirates began to make the long voyage around the Cape of Good Hope beginning what later became known as the "Pirate Round."

On the way, they discovered that Madagascar made an excellent place to obtain fresh water and supplies. It had many fine harbors and anchorages. It was unclaimed by any European powers and there was no central native authority. The local tribes were all too busy fighting each other to cause much of a problem for the superior firepower of the invading pirates. Accordingly, the pirates started using Madagascar as a safe haven from which they would sail and prey on the rich trade of the Indian Ocean, the Red Sea and the Persian Gulf.

The large number of pirates who settled down in Madagascar is noteworthy. Particularly in the area of sheltered waters of Antongil Bay and the off-shore Island of St. Marie (Nosy Boraha) on the northeast coast. They created alliances with the native chieftains, intermingling with the local population, creating a mulatto race (called Zana-Malata) and often became local chieftains. A report from 1712 estimates the number of pirates on St. Marie at about 400 compared to a local population of about 600.[145] As described by Chris Rule (op. cit.): "Life on Madagascar is carried out at a different pace and men take more than one wife. Soon, they have as many as four or five wives and following on from that, many, many children. In the space of a few years, an ex-pirate could have himself his own tribe—wives, children, dependents and followers."

The majority of the pirates who settled in Madagascar were English, French or American. The trade relationship between these settling pirates and American merchants was proving to be extraordinarily profitable. To quote Chris Rule (op. cit.) again: "England had imposed many trade restrictions. Merchants and investment syndicates in America found it profitable to send ships, whose Captains were often 'ex'-pirates under the guise of licensed slave traders with goods that the Madagascan pirates needed. On their return, the ships would deliver slaves for the colonies along with gold, silver, jewels and silks—pirate booty captured from the Moors' trading vessels bought at low prices. The precious jewels and metals were sent on to Europe, bypassing England along the way, to be traded for expensive manufactured goods, which were then traded on to the colonies. The profits were great all round which meant it was worth the risks of the long voyage and possible capture by the British trade officials."

Piracy in Madagascar ebbed with the British royal decree of clemency of 1698. With the end of the War of the Grand Alliance (King William's war, 1689-97) a royal squadron of four men-of-war—the *Angelsea, Harwich, Hastings* and *Lizard*—sailed in January 1699 from Portsmouth for Madagascar under Commodore Warren. They carried four royal commissioners with them. They offered a pardon to all pirates who voluntary surrendered from which Every and Kidd were excepted. After calling at St. Augustine Bay in April 1699, where several pirates made their submission, the squadron reached Tellichery in India in November. As it came to its anchorage, Warren died and was succeeded in the command by Commodore Littleton. In

Figure 65. Pirate and Madagascan maiden

May 1700, Littleton was back on the Madagascar coast remaining in communication with the pirates for eight months before returning to England.

In May 1703, Commodore Richards sailed from London with the *Severn* and *Scarborough* to put an end to piracy. He captured the pirates David Williams and John Pro who managed to escape in Anjouan. Richards died there. The command was passed on to Captain Harland, and the ships were back in England in October 1705. Many

Pirate Utopia

The most remarkable pirate utopia was the attempt around 1690 to establish a pirate-state in Diego-Suarez in the North of Madagascar called Libertalia, under the leadership of Captain James Misson. Misson was French and had served in the Antilles on the war ship *La Victoire*. After putting in at Rome, he befriended the defrocked monk Caraccioli who converted him and most members of the crew to a type of atheist communism. The *Victoire* then sailed back to the Antilles and engaged in a battle close to Martinique with the English vessel *Winchelsea*. The *Winchelsea* was sunk but not without putting up a good fight in which all French officers were killed. The 200 large crew designated Misson as captain and Caraccioli as first mate, collectivized the booty of the vessel and decided to manage everything in common. They captured a few vessels in the Antilles and then crossed the Atlantic. Along the Gold Coast (Ghana), they captured a Dutch slaver, the *Nieuwstadt*, liberated the 17 slaves and brought them on board. There were other captures, resulting in further French, Dutch, English and liberated African slaves joining the crew. They kept one of the captured ships, renamed her the *Bijoux*, and made Caraccioli captain of the prize. After refreshing in the south of Madagascar they put in at Anjouan island (Comoros). They were well received by the Queen and helped out in the fight against her enemy, the Sultan of the nearby island of Moheli. Eventually, Misson married the Queen's sister and Caraccioli did likewise with the Prime Minister's daughter. Some crew members followed their lead, married locally and settled in Anjouan after having received their share of the bounty.

Wishing to pursue his dream, Misson sailed the *Victoire* and the *Bijoux* down the Mozambique Channel. They captured a Portuguese vessel in a fierce fight during which Caraccioli lost his left leg. This forced them to return to Anjouan to heal the wounded and repair the vessels. Two months later, having rounded up his crew, women and about 300 Anjouanese on loan from the Queen, Misson set off for Madagascar. He chose the Bay of Diego-Suarez for it was blessed by fertile soil, fresh water and friendly inhabitants and easily defended given its narrow entranceway. There they set up the 'maritime republic' Libertalia and called themselves Liberi, creating their own language and farming and raising cattle in common. Meanwhile, they launched piracy expeditions. After capturing a Portuguese vessel off the coast of Zanguebar, they met the pirate Thomas Tew, invited him to join them, and named him Admiral of the fleet. Alongside the Angola coast, the *Bijoux* commanded by Tew and his crew captured an English slaver with 240 men, women, and children on board and brought them to Libertalia. Later they built two 80-ton sloops, calling them *Childhood* and *Liberty*, that were used on a four-month trip around Madagascar charting the coast. They then sent an expedition to the Red

Sea and captured a richly laden Moorish ship that was brought back to Libertalia. The ship was taken to pieces for the timber and the guns used in defending the mouth of the harbor. By then, they had cleared and cultivated a large area of land and had 300 heads of cattle. One morning, one of the pirate sloops came back into the harbor chased by five tall Portuguese ships. The well-defended settlement succeeded in sinking two of the ships with many men drowning and in capturing another while two were able to flee in crippled condition.

Later, Tew took the *Victoire* on a cruise in search of recruits. He first called on 23 of his former crew members who had established a small colony a few years earlier on St. Marie Island. Late that afternoon, a violent storm came up suddenly with so high a sea that Tew could not go out to his ship. The storm increased and in less than two hours the *Victoire* parted her cables and was driven ashore, with the whole crew drowning in sight of Tew, who could not help them. Tew stayed with his former shipmates, trying to find a way to return to Libertalia. About four months later, while going to the beach in the morning, to his surprise Tew saw the two Libertalia sloops anchored, from which a boat was swung out with six rowers including Misson. Misson told Tew that all their dreams of happiness had evaporated, as a large group of natives, taking advantage of the absence of the *Victoire* and the *Bijoux,* had suddenly attacked and massacred most of the colony without mercy. Only Misson and 45 of his men succeeded in escaping with the two sloops with some of the gold and diamonds. Caraccioli had been killed.

Tew suggested that they both sail to America, where they could live comfortably from Libertalia's remaining booty. However, Misson wished to return to France. They waited for a week for the *Bijoux* to come back. When she did not appear they divided up the booty and sailed off. Tew took one of the sloops with 30 men and Misson the other with 15 men. Off Infantes, before reaching the Cape, a great storm overtook the two vessels and sank Misson's sloop within a musket shot of Tew who could give no assistance. Tew returned to Newport, Rhode Island, where his crew took their share of the treasure and quietly dispersed. Unable to obtain an amnesty in Rhode Island, he went back to the Indian Ocean on the *Amity* and joined Henry Avery. In September 1695, Avery, Tew and some other pirates joined forces to attack the rich fleet of the Great Moghol in the Red Sea. Tew was killed by a cannon shot during the battle.

Sources

[1] Captain Johnson, Ch., op. cit, vol. II, pp. 1-48 and 81-109; Pouillaude, D., "Le Grand Livre des Aventuriers des Mers, Pirates, flibustiers, boucaniers, corsaires...," Orphie, 2005, pp.60-66; Bastions Pirates, Do or Die, op. cit.

pirates surrendered to the conditions of the pardon. Coupled with regular patrolling of British warships around the Indian Ocean to hunt the pirates down and subdue them, piracy in and around Madagascar became less frequent. Most pirates settled down, enjoying their booty, taking wives and trading in slaves. A new wave of pirates came to Madagascar around 1719 after the new governor of the Bahamas, Woodes Rogers, subdued piracy there. New acts of piracy along the Malabar Coast of India and in the Red Sea ensued.

In February 1721, a squadron of warships consisting of the *Lyon, Salisbury, Exeter* and *Shoreham* and the *Grantham* acting as a storeship under the command of Commodore Thomas Matthews left England with orders to obliterate the pirates in the Indian Ocean. Arriving at St. Augustine Bay with the *Lyon* and the *Shoreham* ahead of the *Salisbury* and the *Exeter*, that had put into Lisbon for repairs, they found no pirate ships there at the time. Matthews carelessly left a letter for Captain Cockburn of the *Salisbury* in which a number of particulars were given of the squadron. He proceeded to India. Soon after his departure, the pirates Taylor and La Buse got hold of the letter and news of the squadron spread like wildfire. The pirates vanished, burning their ships or sending them away, abandoned their houses in St. Marie and settled in small groups on Madagascar's mainland. Some of the pirates returned to the Caribbean. The English squadron searched most of Madagascar, especially St. Marie Island, but found no pirates. They proceeded to Bombay before returning to England without having accomplished anything.[146] While the mission was a failure it had the indirect effect of dampening piracy in Madagascar. Most of the remaining pirates eventually returned to St. Marie but took retirement enjoying the Madagascan ladies, raised families and lived the life of a trader having gone "native." Many were buried in the pirates' graveyard in St. Marie.

CYGNET[147] (1689)

Captain Charles Swan was a privateer in Central America. As an experienced Captain he commanded a ship during Sir Henry Morgan's 1671 sack of Panama. He then roamed north and south on islands and mainland. In 1682, he brought a merchant vessel from Jamaica to London. There he befriended Basil Ringrose and the two men persuaded some London merchants to outfit a ship for trade along the western coast of South America. The 180-ton, sixteen-gun

Figure 66. The pirate cemetery in St. Marie

Figure 67. The pirate cemetery in St. Marie

Cygnet under the command of Captain Swan was chosen. He had thirty-six men of whom Ringrose was one. She sailed from the Downs on October 1, 1683 with a cargo worth £5,000 (equivalent to $12.5 million today).

As a trading voyage it was a disaster. Instead, it became an unending interminable cruise. While ravaging the coast of Peru in late 1683 Swan met up with the pirates Edward Davis and Peter Harris. When Swan's crew saw that Harris' pirates had handfuls of gold in their pockets, they forced him to cross the fine line between semi-legitimate trade and piracy. Captain Swan proceeded to fire off several letters to the owners of his ship *Cygnet* in London begging them to intercede with James II of England for his pardon—even as he joined forces with the two other pirates to loot his way round the coast of South America. However, he had no great success. At Valdivia (Chile) in March 1684 they were driven off by the Spaniards despite a flag of truce with two men killed. Ringrose and one other were the only ones of the landing party to escape unhurt. To quote Dampier: "Captain Swan began to repent that ever he took this voyage in hand and he did never affect Master Ringrose afterwards...for Mr. Ringrose being the proposer of this voyage did demonstrate the thing being very feasible in England which now Captain Swan found to be difficult."

The three pirates took part in the attack on Payta in Northwestern Peru in 1684. They petulantly burnt down the town after nothing worth stealing was found. Then they tried to trade in the gulf of Nicoya in Costa Rica with equally disastrous results. In February 1685 they sailed their ships to the Gulf of Panama. They waited there in the hope of intercepting Peruvian treasure ships. They were soon joined by Captain Townley who was in command of several hundred French buccaneers. Because of general failure in legitimate trading and the desertion of many of his crew, Swan decided that the *Cygnet* should team up with Townley. This constituted a force of ten buccaneer vessels. However, in June the buccaneers were defeated by Spanish warships and the French and English contingents split up. The English raided Nicaragua with little success. On August 25, 1685 Swan separated from his confederates Peter Harris and Edward Davis while taking William Dampier on board.

William Dampier was an English buccaneer, navigator, hydrographer and naturalist who eventually circled the globe three times and gave the English language many new words. Having become an orphan early, he was placed with the master of a ship at Weymouth and made several voyages in merchantmen. He served in 1673 in the

Dutch war and in 1674 became an under-manager of a Jamaica estate for a short while. Afterwards he engaged in the coasting trade, made two voyages to the Bay of Campeachy in the Gulf of Mexico and remained for some time with the logwood-cutters. He varied this occupation with buccaneering. After serving with a privateering expedition in the Spanish Main he went to Virginia and engaged with Captain John Cook on the *Batchelor Delight* for a privateering voyage against the Spaniards in the South Seas. They sailed in August 1683, touched at the Guinea coast and proceeded around Cape Horn into the Pacific, cruising along Chile and Peru. They took some prizes and proceeded to the Galapagos Islands[148] and to Cape Blanco in Mexico. While they lay here Captain Cook died and the command devolved on Captain Davis. He returned to South America and joined forces with Swan's *Cygnet*.

The *Cygnet*, with Dampier on board, proceeded north up the coast of Mexico looking for a Manila galleon. Swan tried to capture a Peruvian treasure ship at Acapulco but failed because the vessel was protected by a battery of Spanish guns on the mainland. He also missed the Manila galleon. He continued north along the northern parts of Mexico as far as southern California in search of gold and silver mines. This ended up in failure and the *Cygnet* returned to Mexico. On February 19, 1686 Swan and his men landed at the mouth of the Rio Grande de Santiago in Mexico opposite the Tres Marías Islands seeking provisions. They captured the small town of Sentispac (Santa Pecaque) fifteen miles inland without resistance. However, while they were transferring supplies of maize onto horses to take to their canoes in the river a large body of Spaniards ambushed the English party. They killed Basil Ringrose and 50 of the buccaneers, a quarter of Swan's entire force.

As the expedition failed to find success on the Mexican coast, Captain Swan proposed to run across the Pacific and return by the East Indies. On March 31, 1686 the *Cygnet* set out across the Pacific with a hundred men on board, together with a bark commanded by Captain Josiah Teat with 50 men and some slaves. They steered due west for the Ladrones (Northern Marianas) Islands. However, due to the failure of the assault on Santa Pecaque provisions were short. As the food began to run out the crew blamed those in charge and plotted to eat their officers. Swan is reported to have remarked that the lean William Dampier would have made them a poor meal; the Captain himself was a remarkably fat man. Fortunately, they reached Guam in only 52 days without resorting to cannibalism with three days rations to spare.

In early June, the *Cygnet* and the bark sailed to the Sultanate of Mindanao (Philippines). Swan's men spent their money there on women and wine in a profligate manner. So confident was Captain Swan in the good intentions of the local ruler Raja Laut, that he carried his vessel over the bar into the river. They soon discovered that her bottom was perforated by the teredo and that a short time before a Dutch vessel was entirely destroyed by them in less than two months. Raja Laut's true intentions to seize the *Cygnet*'s cargo and guns became clear.

The crew ripped off the worm-eaten planks, put on new, and sheathed and tallowed the ship's bottom before getting her back over the bar. They pressed Captain Swan to be gone but he was living on shore and refused to make a decision. The mutiny was brought to a head by the discovery of the Captain's journal in which he inveighed against the crew. On January 13, 1687, the mutineers weighed anchor and were standing out to sea. A last ditch effort was made to convince Captain Swan to come on board but he refused the offer. They left him and about 36 other crew members stranded on Mindanao. Reportedly, Captain Swan, while attempting to get on board a Dutch ship later that year was tousled by the local population and drowned.

The *Cygnet* left Mindanao on January 14, 1687 directing her course to Manila. John Read was chosen as Captain and Josiah Teat as master. Dampier was also kept on board against his will. They captured a few small Spanish vessels before hiding in Pulo Condore (Con Son, off the south coast of Southern Vietnam) where they careened the ship and made a new suit of sails. On April 21 they overtook a Chinese junk and anchored next off the island of Sillabar (off the coast of Sumatra) where Dampier made an unsuccessful attempt to escape the pirates. Next, they sailed to the island of Saint John on the south coast of Canton in China. They were pushed by a storm to the Pescadores Island (in the Straights of Taiwan) making the group on July 20. Sailing on the *Cygnet* touched at one of the Bashee Islands and proceeded to sail south round the eastern side of the Philippines to the Celebes islands. They brought up off Bouton Island in the southeast corner of the Celebes on December 15. From there the *Cygnet* steered across for New Holland (Australia).

The *Cygnet* sighted New Holland on January 4, 1688. Captain Read found a pretty deep bay with abundant islands in it and a very good place to anchor in or to haul ashore. About a league to the eastward of that point she anchored on January 5, 1688 two miles from shore. The bay in which the *Cygnet* anchored is still called Cygnet Bay today; it is situated in the northwest corner of King's Sound in the district of West Kimberley in Western Australia. "That point" is named Swan Point while a rock is called Dampier's Monument to commemorate the buccaneer's visit, the first landing of an Englishman to explore Australia.

The ship remained in Cygnet Bay until March 12, 1688. During that time the vessel was hove down and repaired. Dampier's observations on the aboriginal inhabitants during his stay is summed up in his description of the natives whom he saw. Who were, he says, "The most miserable people in the world." He gives an accurate description of the country so far as he saw it and asserted that "New Holland is a very large tract of land. It is not yet determined whether it is an island or a main continent; but I am certain that it joins neither Asia, Africa nor America."

The *Cygnet* then steered northwest for the Nicobar Islands (east of the Indian mainland), where she was again careened in order for her to sail faster. After the ship floated again Dampier insisted on being landed and at last the Captain agreed. He was set ashore with three other men, traveled to Sumatra by canoe and worked his way to England which he reached in 1691. The *Cygnet* made for Ceylon and was driven ashore on the coast of India. Captain Read, assisted by Teat as first mate, scoured the Bay of Bengal and took several Spanish and Portuguese prizes. Pushed by the prevailing winds the *Cygnet* ended up on Madagascar's east coast where the crew provided assistance to a local king in a battle against his neighbor. Afterwards Read left with five or six other crew members on the American slave ship *New York*.

Teat took over as commander of the *Cygnet,* renamed her *Little England,* rounded the south of Madagascar and met a brigantine commanded by Captain Knight at St. Augustine Bay. They sailed in consort to the coast of Sofala where they plundered a Portuguese ship of its cargo of elephant teeth and amber. They touched at Moheli (Comoros) where they met another Portuguese vessel, killed her commander, plundered the ship of all her cargo of cloth and cut all her masts. Both ships then proceeded to Anjouan island. On June 9, 1689 they met the survivors of the *Anne* East Indiaman who had sailed from Morondava with the sloop and the longboat (see story above). Eight crew members of the *Anne* shipped themselves on board the pirate ships. Having been informed by the natives that the East Indiaman *Chandois* was expected about this time the two ships sailed on

June 11 directing their course directly northwards to the Red Sea. The run was not very successful, however. In need of provisions they returned to St. Augustine Bay. Being very much worm eaten, the *Cygnet* sank in late 1689 in the Bay of St. Augustine while at anchor close to six years after her departure from London.

JOHN AND REBECCA[149] (1697)

Captain John Hoar was from Rhode Island. Under a privateer's commission issued by the English authority in Jamaica he captured a 200-ton, 14 gun French ship in the Caribbean in 1694. He brought the prize back to Rhode Island. Abraham Samuel, a mulatto son of a Martinique planter and a black slave, being on the run had joined her crew. A new Admiralty Court, established under Captain Hoar's petition, duly declared the ship and her cargo legally that of Captain Hoar. He renamed her *John and Rebecca* and fitted her out as a privateer. Having received a privateering commission from Governor Benjamin Fletcher of New York, Hoar left Boston in December 1695 and sailed for richer prey in the Indian Ocean. He and his fellow pirates eventually elected Abraham Samuel as quartermaster. They put in at St. Marie Island in April 1696 on the way.

Captain Hoar next joined the Dutch pirate Dirk Chivers, Captain of the *Soldado* and they seized several Indian and European ships in the Red Sea.[150] Hoar then sailed alone to the Persian Gulf where he captured a large Indian ship loaded with cloth near Surat. Now possessing an important booty the pirates decided to retire and the *John and Rebecca* sailed for St. Marie Island anchoring in February 1697. Unbeknownst to them, the Malagasy had rebelled against Adam Baldridge, a retired pirate who became the go-between for the pirates and New York merchants who bought their booty. Captain Hoar and a number of pirates died in the uprising. Samuel and others escaped and the command of the *John and Rebecca* passed on to Abraham Samuel. They set sail for New York, but the ship, which was in a sorry state, sank after hitting a reef near Fort Dauphin in 1697.

While bathing one day, local people recognized specific birthmarks indicating that Samuel could be the long departed son of the queen. As the old queen was still alive, he was brought to her and she recognized him as her son whom she had conceived with a Frenchman long before and who had left with his father as a child. Captain Samuel accepted this fallacy and became the sole heir to the throne.

He fought and subdued the local ruler with the help of some of the shipwrecked pirates who had remained with him and became the King of the Anosy region ("Tolinor Rex"). He always kept close bonds with his pirate brethren, authorizing and even helping them out in the plunder of merchant vessels that came to Anosy to trade with him.

ADVENTURE GALLEY[151] (1698)

The *Adventure Galley* was a 287-ton galley, mounting 34 guns and carrying 23 pairs of oars captained by William Kidd. She was built in haste in London in 1695 after Kidd had accepted a mission to capture French ships and pirates in the Indian Ocean. Kidd was born in Scotland, took to the sea as youth, emigrated to America and by 1689 was Captain of the *Blessed William,* a privateer in the king's service sailing against the French in the West Indies.

Shortly after attacking a French sugar plantation at Mariegalante, Kidd was ordered to join another squadron to attack French warships in a sea battle. His crew refused, arguing that they were paid out of the proceeds from looting and pillaging but would get nothing for taking part in such a hazardous mission. Kidd confronted them and with the matter still unresolved went ashore on Nevis in the West Indies. In the middle of that night of February 1690, his angry crew, led by Robert Culliford, cut cable and stole the *Blessed William* along with Kidd's booty in the hold. Angry and destitute Kidd was given another ship, a sloop named *Antigua,* and crew by the Governor. He chased his duplicitous men to New York. In 1691 he married a wealthy widow and became a land-owner in the British colony of New York and a confidant of colonial governor Colonel Benjamin Fletcher.

Meanwhile, the King of England William III wished to rein in the rampant piracy in the Indian Ocean, Red Sea and Persian Gulf and the ensuing loss of East India Company vessels month after month. Yet war with France prevented the Royal Navy from sending warships to chase the pirates down. In early 1695, Kidd chanced upon a meeting in New York with Colonel Robert Livingston, a fellow Scotsman, ambitious entrepreneur and confidant of King William III. Aware of the problems plaguing the Crown with regard to piracy, Livingston concocted a scheme for striking at the pirates and making a profit at the same time. His proposal involved outfitting a specially built privateer to seek out the pirates, bring them to justice and confiscate their booty. Captain Kidd sailed to London in the summer of 1695 aboard his sloop *Antigua* on a trading run and to meet up again with Livingston to put the final touches to their plan. Livingston had found financial backers to outfit the ship. They put pressure on Kidd, who knew the ways of pirates, to accept the privateer appointment. Eventually Kidd accepted.

Figure 68. Seventeenth century portrait of Captain William Kidd (believed to have been painted from a sketch drawn in Newgate whilst Kidd was awaiting trial)

The *Adventure Galley* was launched at Deptford on the Thames River in December 1695. Kidd hand-picked his crew of 70 from married seamen with families in England whom he believed would not turn to piracy. He sailed for New York on February 27, 1696, planning to recruit a further 80 men. He was barely underway when he failed to salute a Royal Navy warship in the Thames estuary. He was stopped by the man-of-war who impressed most of his crew leaving him with barely enough men to manage the ship. He sailed down the English Channel and called in to Plymouth where he recruited more crew for his voyage across the Atlantic. He made landfall in New York in July and recruited more crewmen of dubious quality.

He sailed for the east coast of Africa in September of 1696, rounded the Cape of Good Hope in December, put in at Madagascar in February 1697 and proceeded to the Comoros islands. While careening the ship on the island of Moheli, 50 of his men fell ill and died in the space of a week. With provisions beginning to dwindle and no penny earned so far, the crew began openly advocating piracy by attacking any ship. Kidd was under pressure. He could not afford to lose another crew and risk total failure. With his mission hopelessly off course, Kidd tried to uphold the terms of his commission

Figure 69. The *Adventure Galley* (drawing by John Batchelor)

as best as he could but faced with the constant threat of mutiny he opted for the compromise that branded him a pirate. On April 27, 1697, they set sail northward toward the Red Sea. After anchoring off the island of Perim at the mouth of the Red Sea in July Kidd tried to attack a convoy of Moorish vessels but was driven away under fire by an escort vessel, the East Indiaman *Sceptre* commanded by Captain Edward Barlow. Kidd's prestige waned further in the eyes of his crew.

In August 1697, Kidd seized a small brigantine *Mary* near Janjira. He tortured the Indian sailors and impressed the British Captain into acting as pilot for several months. He then continued south following the coast past Kárwár and Calcutta, engaging two Portuguese warships in September and stopped at the Laccadive Islands for repairs where Kidd's men mistreated the islanders. In early November, he halted an English cargo vessel named *Loyal Captain* sailing northward along the Malabar Coast but let her go. His crew was infuriated, but Kidd faced them down. The mutiny faded out. However the ill temper of the crew did not. Some two weeks later, Kidd became involved in an argument with William Moore, the Galley's gunner, over whether or not to attack a Dutch vessel they happened upon. Moore wanted to attack the passing ship and during the altercation Kidd seized an ironbound bucket and crashed it against Moore's head. Moore died the next day of a fractured skull.

At the end of November 1697, the *Rouparelle* (sometimes referred to as the *Maiden*), hoisting a French flag and bound for Surat with a cargo of cotton, quilts and sugar was stopped by Kidd and his crew. Believing that he had captured a legitimate prize, Kidd set the Moorish crew free and sold the cargo on shore for cash and gold which he passed out to his crew. He renamed the ship *November* and took her along as a prize.[152] As it turned out, the ship was actually Indian-owned and legally Kidd had now committed piracy. In December, Kidd seized two other small vessels but the take amounted to very little.

On January 30, 1698, the *Adventure Galley* in heavy seas off the Indian coast north of Cochin encountered a 500-ton merchantman by the name of *Quedagh Merchant*. She was Armenian-owned and captained by an Englishman by the name of Wright. Her rich cargo contained silks, muslins, sugar, iron, saltpeter, guns and gold coins. Kidd pursued her for four hours before finally drawing alongside and firing a shot across her bow. He claimed her as a prize, sold some of her cargo on shore for £10,000 (worth $25 million today) and divided the money amongst his crew.

Figure 70. Captain Kidd hurling a bucket at William Moore's head

Figure 71. An engraving of the *Adventure Galley* attacking the *Quedagh Merchant*

Figure 72. William Kidd (Painting by Howard Pyle, 1902)

Enriched, Kidd then sailed for Madagascar with both the *November* and the *Quedagh Merchant*. Along the way they seized a Portuguese vessel but were unsuccessful in their pursuit of two East India Company ships, the *Dorill* and the *Sedgwick*. On April 1, 1698, the *Adventure Galley* landed at St. Marie's island to take a well deserved rest. The pirate ship *Mocha* lay at anchor in the bay. Kidd got the shocking news that her commander was the very same wicked Robert Culliford who was involved in stealing his ship in Nevis eight years earlier. According to the deposition at his trial, Kidd wanted to get even with Culliford but his men refused to seize the *Mocha*. Instead, when the *Quedagh Merchant* caught them up a month later, they insisted on splitting the booty. All but 13 of his crew deserted and joined Culliford. They scuttled the *November* and then stripped the *Adventure Galley* and *Quedagh Merchant* of guns, small arms, powder, shot, anchors and cables, burned Kidd's logbook and threatened him with murder. Kidd initially barricaded himself in his cabin but eventually surrendered to Culliford saving not only his own life but that of the men who remained loyal to him. The *Adventure Galley* now rested on a sandbar in the shallows, leaking and half full of water. Kidd burned her, fitting out the *Quedagh Merchant* with what he could salvage from his stricken vessel.[153]

Kidd then spent the next five months scrounging up a crew for the voyage home while waiting for the northeast monsoons which could blow him around the Cape of Good Hope.

He departed aboard the *Quedagh Merchant* which he had renamed the *Adventure Prize* on November 15, 1698 for the return voyage. When he reached the Caribbean island of Anguilla in April 1699, Kidd received the news of his new status of pirate and a much-wanted man. Believing in his innocence and in the men who had hired him he sailed for New York. However, before meeting Governor Bellomont he took the precaution of burying various items of his booty amongst several scattered caches along the banks and on various islands in the Hudson River. If taken into custody the booty could serve as a bargaining chip. He made landfall in Boston in July 1699, was arrested and sent to London. His treasure was tracked down, recovered and shipped back to London. He was tried in May 1701, sentenced to death for piracy and murder and hanged on the Wapping waterfront. Kidd's body was tarred and bound in chains, with his head encased in an iron frame and left to hang for years at Tilbury Point on the Thames—to serve as a warning for would-be pirates.

In 1999 and 2000, Barry Clifford and his Project Team completed three major expeditions to Saint Marie Island as a *Discovery Channel* "Quest" initiative, where five shipwreck sites were discovered, believed to be the *Adventure Galley* and the *Flying Dragon* commanded by the pirate William Condon. The other three shipwreck sites have been tentatively identified as the *Rouparelle,* the *Mocha* (Captain Culliford) and the *New Soldado* (Captain Chivers).

Figure 73. Kidd's body hanging at Tilbury Point

ROUPARELLE (NOVEMBER) (1698)

In mid-1696, Captain John Hoar on the *John and Rebecca* cruised with Captain Dirk Chivers on the *Soldado* and seized several European and Indian ships in the Red Sea. One of these was the 150-ton East Indiaman *Rouparelle*. They let her go and she later passed under Indian ownership. In late November 1697 the *Rouparelle*, while bound for Surat with a cargo of cotton, quilts, and sugar was captured again. This time it was by William Kidd on the *Adventure Galley*. Kidd renamed her *November* and took her as prize to Saint Marie Island. In May 1698 she was taken from Kidd and sank by Captain Culliford next to the *Mocha* at Careen Key at Saint Marie's so the *Mocha* could be careened.

NEW SOLDADO[154] (1699)

The history of the *New Soldado* is intimately linked with that of the Dutch pirate Dirk Chivers. In early 1694, Chivers signed aboard the *Portsmouth Adventure* under Captain Joseph Farrell. The ship was leaving Rhode Island for the Red Sea. Once there Farrell helped Henry Avery capture two ships rich in bounty around June 1695. While returning to Rhode Island the ship was wrecked on the Island of Mayotte (Comoros). Farrell was rescued by Avery and continued on with him but Chivers stayed behind on the island. At the end of 1695 he joined the 200-ton, 28 gun ship *Algerine* that had sailed in 1693 under the command of Captain Robert Glover from New England to Madagascar. A few months later, Dirk Chivers lead a mutiny and took over the *Algerine*. After becoming Captain he renamed her *Soldado*.

In mid-1696, Chivers sailed in consort with Captain John Hoar of the *John and Rebecca*. In the Red Sea they took two English East Indiamen on a voyage to Surat with a ship-load of Arab horses. As neither Captain wanted to take possession of the large slow moving cargo vessels they decided to ransom them to their owners. While sailing to Aden, the constant complaining of Captain Sawbridge, a captive from the seized vessels, provoked Chiver's crew into sewing the man's lips with a sail needle to silence him. They kept him thus several hours with his hands tied behind him. When the Governor of Aden refused to pay the ransom, they set the ships on fire, burning them to the waterline and the horses in them. Sawbridge and his people were carried to Aden and set on shore where he died soon after.

After that, Hoar and Chivers parted and Chivers proceeded to India. In November 1696, he seized four ships at Calcutta and demanded £10,000 (worth $25 million today) for their safe return. Not swayed by Chivers' threat, the Governor sent 10 Indian ships to attack him and reclaim his prizes. When they sailed into the harbour, Chivers knew he was outgunned and fled taking moderate damage in his escape. He chose to head for St. Marie to make repairs with the *Soldado* arriving there in the summer of 1697.

After repairing the *Soldado,* Chivers put back to sea and in April 1698 captured the English trading vessel *Sedgwick* off Cape Comorin. The cargo of the *Sedgwick* not being to Chivers' liking he let her go taking only her sails and cordage. Chivers next joined forces with Captain Robert Culliford of the *Mocha* and Captain Nathaniel North of the *Pelican* to capture a 600-ton Moor pilgrim vessel, the *Great Mohammed,* in the Red Sea—the biggest capture in pirate history in the East Indies. The *Soldado* and the *Mocha* were first to attack her. The crew and passengers numbering about 1,000 begged for mercy before the *Pelican* arrived to participate in the capture. The *Soldado* had grown unseaworthy and was abandoned on the Malabar coast. Chivers transferred to the *Great Mohammed* and renamed her *New Soldado.* He then returned to St. Marie.

The British royal decree of clemency was adopted in 1698. The next year Commodore Warren was sailing to Madagascar with a squadron on four ships and this Act of Grace for practicing pirates. In St. Marie's harbor were the *New Soldado* (Dirk Chivers), *Mocha* (Robert Culliford) and *Dolphin* (Samuel Inless) all of which were sunk with other ships at the mouth of the harbor to blockade the approaching squadron. Some other Captains were quick enough and fled the harbor. Chivers accepted the royal pardon and returned home on the *Vine.*

MOCHA[155] (1699)

The *Mocha* was a 150-ton frigate put in service of the HEIC in 1694 under the command of Captain Edgecumbe. While in Bombay in 1696, Edgecumbe enlisted about 20 men headed by Ralph Stout. In 1692, Stout was arrested in India with 20 other pirates while serving under Captain James Kelley. They spent four years in an Indian prison before escaping and stealing a boat. They made their way to Bombay where they joined the *Mocha*. Barely eight days after leaving port on the way from Bombay to China the crew of the *Mocha* mutinied off the coast of Acheen. In the mêlée Captain Edgecumbe was killed. Twenty seven loyal officers and men managed to escape and reach shore in the *Mocha*'s pinnace. The men elected Ralph Stout as their new Captain and discovered £19,000 worth of booty on board (or $47 million today).[156] Stout steered the newly captured ship to the Mergui Archipelago (southeast of Burma) where they encountered and captured a merchant ship. In early 1697 they sailed to the Nicobar Islands where they rescued the marooned Robert Culliford.

Robert Culliford had arrived in the Indian Ocean in December 1690 on the *Blessed William* which was stolen from William Kidd in Nevis under the command of the American Captain William May. After setting a base in Madagascar, they went on the hunt on the Indian Coast. Arriving at Nicobar Islands, May and Culliford jumped ship and returned to New York where May was given the command of the *Pearl* and Culliford became quartermaster. Together they sailed for Masore (India). Soon after arriving in October 1694, Culliford left the ship and signed aboard the *Josiah* ketch from Bombay as a gunner. In June 1696, while at anchor in Madras Culliford led a mutiny taking advantage of the commander being on shore and ran away with the ship. The piratical career of the *Josiah* did not last long though. Making first for the Nicobar Islands, the crew flocked on shore and were soon involved in quarrels with the natives. They left two men on board and one was James Cruffe, the armourer, who was forced to join them against his will. The other man a lukewarm pirate and Cruffe prevailed on him to join in an attempt to carry off the ship. They cut the cable and by great good fortune having no knowledge of navigation succeeded in carrying the ship into Acheen. Culliford was marooned on the island with the other men.

In April 1697, shortly after rescuing Culliford and his men, Stout sailed the *Mocha* to Laccadive Islands (southwest of India) where he was murdered. According to one account, he was put to death by his comrades for trying to desert them; according to another account, he was slain by Malay seamen visiting the islands. His place was taken by Culliford.[157] He sailed to the Strait of Malacca and pursued the British ship the *Dorrill,* China-bound from Madras. When the *Mocha* closed in, the *Dorrill* opened fire and sheared the *Mocha*'s main mast. After a hot engagement of three hours and severely damaged Culliford had to retreat and decided to sail for Sainte Marie's Island to undertake repairs. En route, he captured several ships and his booty came to £2,000 ($5 million today). While in Sainte Marie he met his old one time friend William Kidd. Former grievances forgiven they apparently enjoyed each others company.

In June 1698, Culliford left Saint Marie after having increased *Mocha*'s guns to 40. Most of Kidd's crew had enlisted with him. He joined forces with Dirk Chivers on the *Soldado* in capturing the *Great Mohammed* in the Red Sea in September 1698. In February 1699, while returning to his base at Saint Marie he captured a French ship. *Mocha* was sunk with the *Soldado* and *Dolphin* at Saint Marie's Island to blockade the harbor against the British Squadron of 1699. Culliford accepted the royal pardon and went to London where he was arrested and tried for piracy but saved from hanging because his affidavit was needed for the coming trial of Samuel Burgess. After his court case Culliford seems to have disappeared from record.

BEDFORD (1697) AND DOLPHIN[158] (1699)

The *Dolphin* was an Arab ship captured in 1698 by Nathaniel North. During the nine-year war (1688-97) North was a crew member aboard a French privateer ship in the Caribbean. He refused to join the British Royal Navy around 1696 and went with the privateer that captured the French merchantman, the Bristol built, 18-gun *Pelican* off Newfoundland. After putting in at Jamaica, they got a commission to cruise southwards and put to sea again bound for Madagascar under Captain de Grammont. The *Pelican* arrived in St. Augustine Bay in July 1697. As it was too late to sail to the East-Indies they toured and pillaged some towns in the Comoros islands before returning to St. Augustine. Some of the men were stricken by illness. North was elected quartermaster. They also picked up some stranded pirates including Captain David Williams.

David Williams, originally from Wales, was abandoned ashore by his merchant ship *Mary* in the late 1680s on Madagascar's east coast while he was searching for water. He spent close to ten years living with a number of local kings, including several years with the King of Matitanana and then with King Andrianampoina. He was very much esteemed for his bravery in battles. In mid 1697, a pirate galley the *Bedford* commanded by Achen Jones, a Welshman, anchored in Matitanana and Williams was permitted to enter her. They went to St. Augustine bay. While careening, by carelessness they broke her back and lost her. The crew lived in St. Augustine until the arrival of the *Pelican.*

In September 1698, the *Pelican* joined the company of the *Soldado* (Chivers) and the *Mocha* (Culliford), agreed to make an equal division of all prizes and Williams shifted on board of the *Soldado*.[159] About ten days later, they gave chase in the Red Sea to a large Moor ship the *Great Mohammed.* The *Soldado* and the *Mocha* attacked her first and before the *Pelican* could enter a man the Moors called for quarters. The bounty was divided among the crew of the *Soldado* and the *Mocha* who refused to share it with the crew of the *Pelican* on the pretext that they had not participated in the battle. The crew of the *Pelican* protested but was told to be gone or they would sink them. Hence, the *Pelican* was left to pursue her fortunes along the Malabar Coast, eventually capturing three Arab ships. One was armed with 26 cannons and renamed *Dolphin.* Having abandoned the *Pelican* at sea she sailed for Madagascar. Taken in a massive storm off Bourbon, the *Dolphin* lost all her masts but the crew managed to bring her to St. Marie where she was refitted.

In St. Marie they met with Chivers and Culliford and with three American ships that had come to trade with the pirates. One of them was the *Margaret,* belonging to Frederick Phillips of New York and captained by Samuel Burgess.[160] Captain de Grammont and some of his men, tired of the pirate life and returned to America with these ships. The remaining crew elected Samuel Inless, who was living in St. Marie, as the *Dolphin*'s new Captain and North as quartermaster. The *Dolphin* left for the Malacca Straits. After several captures, including a large Danish vessel she came back to Madagascar where the booty was divided among the men. A month later, the *Dolphin* was sunk with other ships at the mouth of St. Marie's harbor to blockade the approaching British squadron.

Nathaniel North did not trust the British. He fled with a few men in the *Dolphin*'s pinnace and sailed for Madagascar's mainland. However, his boat was overturned during a storm, with all hands lost except for North who swam twelve miles to shore. About a year later (1700) North and some remaining crewmembers of the *Dolphin* who had settled in Fenoarivo Atsinanana (Fenerive) were taken in by a small French ship that was captured by the pirate George Booth and North became quartermaster. He took part with Booth and Captain Bowen in the capture of the *Speaker,* originally a French man of war that was captured and refitted as an English merchantman. North sailed aboard the *Speaker* from 1701 to late 1703, first under Booth and, after the latter was captured by Arabs at Zanzibar, with John Bowen. North was elected as Captain of the pirates in Madagascar after Bowen retired in Mauritius late in 1703. The beginning of 1707 marked North as a quartermaster once again sailing under John Halsey aboard the *Charles*.

ALEXANDER[161] (1700)

Thomas Howard, a man who spent all his inheritance, fled to Jamaica where he ran away from his ship and stole a canoe with a motley crew. Howard and his small band seized one small ship after another until in 1698 off the coast of Virginia they captured a 24-gun galley. The crew elected James as her Captain and Howard as her quartermaster. With a man-of-war heaving in sight the pirates crossed the Atlantic bound for the coast of Guinea.

After cruising the coast for some months and plundering more ships they spied a large three-decked Portuguese ship from Brazil mounted with 36 guns. They gave chase and caught up with her. As soon as they were close, the Portuguese Captain fled into the hold but the mate, an Englishman named Rutland, and 30 men fought them for the better part of a forenoon. They eventually had to ask quarters which were given. After finding the Captain hidden in the powder room, they hawled him up and whipped him round the deck for his cowardice. The pirates exchanged their galley for the Portuguese ship, carried her in shore and ripping off her upper deck made her deep wasted and much snugger by cutting down some of her gunnel. They renamed her *Alexander*. Next they sailed down the west coast of Africa, making several prizes. They careened the *Alexander* and took provisions at Cape Lopez (Gabon). Sailing further south, they captured and burned two Portuguese brigantines then shaped their course for Madagascar.

Arriving in Madagascar in early 1700, the *Alexander* was wrecked on a reef near a small island less than ten miles from the mainland most likely Nosy Hao opposite Andavadoaka.[162] The Captain was sick in his bed. The crew carried a great deal of the provisions, water and wood to the small adjacent island to lighten the ship and try to dislodge her from the reef. Howard and eleven seamen who remained aboard, took all the treasure, put it on board the boats and furtively made off for the mainland. Captain James hearing no noise, came on deck and saw them put off. He fired one of the cannons at them but missed. As the sea ebbed the ship lay dry and the crew on the island managed to walk to her. They might even have saved her had they had the boats to carry out an anchor and heave her off. But for the want of boats, they brought everything ashore at the tide of flood on rafts.

As the *Alexander* lay in a quiet place the crew had the opportunity to rip her up and build a vessel out of her wreck. The majority of the crew, of English and Dutch origin, sided together. Considering the small size of the vessel they were building and that their provisions were insufficient for everybody, they forced 36 Portuguese and French to get upon a raft and take their chance with the sea breeze to get to the mainland. They finished a vessel of 60 tons but the day they planned to launch her a pirate brigantine hove in sight and took them on board.

Meanwhile, Howard and his crew had sailed north with the design to round the north end of Madagascar and to go to St. Marie island but strong currents prevented their plans. They waited on an island (Ankoala) in the northwest. There, Howard became the victim: he was marooned while hunting when his companions made off with the boats and the treasure, rounded the north end and settled in Fenerive. A few months later, Captain George Booth came along, having just captured a French ship and recruited Howard's former crew. Howard joined the crew, and with Booth, John Bowen, Thomas White and David Williams captured the *Speaker* in Mahajanga. When Booth was detained in Zanzibar, command of the *Speaker* passed on to Captain Bowen, a Creole pirate from Bermuda. After a run in the Red Sea, the *Speaker* was wrecked in January 1702 in Mauritius.[163]

Howard went back to St Augustine's Bay, settled for a short time and managed to recruit some pirates. They seized the 36-gun ship *Prosperous* that had come into the bay. Howard was elected Captain. He was rejoined by Bowen who had seized the *Speedy Return* and was looking for an alliance in Mayotte around Christmas 1702. In March 1703, Howard and Bowen took the British merchantman *Pembroke* with Captain Woolley off Mayotte Island. In August 1703 they captured two Indian ships off Daman with more than £70,000 in booty (worth $175 million today). While lying off Rajapura, they judged the *Prosperous* and the *Speedy Return* to be unserviceable and burnt them both. They fitted the Indian prize, mounted her with 56 guns, transported both companies aboard her and renamed her *Defiance*. They left Rajapura in late October 1703 to cruise the coast of

Malabar. Following a dispute with Bowen, Howard and 20 of his men landed with their share of the treasure. Howard retired as a wealthy man on the Indian coast and married a local girl. As Johnson puts it: "Being a morose ill natur'd Fellow, and using her ill, he was murder'd by her Relations."

CHARLES, BUFFALO, RISING EAGLE AND DOROTHY (1708) AND NEPTUNE (1709)[164]

The *Charles,* a 200-ton brigantine mounting 10 guns was launched in 1693. She initially sailed under Captain John Churcher who was succeeded by Captain Daniel Plowman. During the War of Spanish Succession (1701-14) she was commanded by Captain John Halsey, an American privateer born in Boston, in the service of Great Britain. When his official commission expired in 1705 the *Charles* was returned to her owners in Boston. Halsey received a new privateer commission to the Red Sea permitting him to rob only Moor ships. Sailing for Madagascar, he first put in at St. Augustine Bay for refreshments and picked up several castaway sailors of the lost *Degrave* formerly under the command of Captain Young.[165] Leaving St. Augustine, he set course for the Red Sea in search of Moorish treasure ships and operated in the Indian Ocean throughout the year 1706 with some success.

In late 1706, Halsey and his gunner were charged with cowardice and relieved of command by their crew after refusing to order an attack on a large Dutch ship in line with their commission prohibiting them from attacking European vessels. The crew, who presumed the ship to be a lone merchantman, went forward with the attack. As they approached their intended victim, the Dutch ship turned to reveal its sixty guns and fired a warning shot towards the *Charles.* This injured the man at helm, unstripped the swivel gun and severely damaged the topsail. The attack by the Dutch caught the crew off guard, causing many to run down between decks. Although suffering damage, the *Charles* was able to escape and Halsey was reinstated as commander.

In February 1707, they seized two coastal traders, the *Buffalo* and a sloop, off the Nicobar Islands. Here a dispute arose among the pirates concerning where to go next. Some of the crew went on board the *Buffalo,* made Rowe Captain and returned to Madagascar. The others ripped up the sloop's deck and mended it with the bottom of the *Charles.* Next, the two ships sailed to the Straits of Malacca but found little success due to the low morale of the crew following the incident with the Dutch ship.

Returning to Madagascar, they anchored at "Hopeful Point" (Foulpointe) where they met with the *Buffalo* and the *Dorothy,* a prize captured by Captain Thomas White.[166] Halsey repaired the *Charles,* recruited more crew including Thomas White, David Williams and Nathaniel North who became quartermaster and set out for the Red Sea. After entering the Red Sea in August 1707 he captured—after a battle lasting close to one hour—two British ships; the *Rising Eagle*[167] and the *Essex* from a squadron of five. His booty was estimated as £50,000 in cash and cargo (worth $125 million today). Halsey discharged the *Essex* and took the *Rising Eagle.* He left North in command of the *Charles* and steered for Madagascar. The *Rising Eagle* and the *Charles* were separated in a storm but both made it to Madagascar. North and Williams fell in with Ambinany Matitanana in the southeast. Finding that the *Charles* was very much worm-eaten and made a great deal of water, North and his crew took all their goods ashore and laid up their vessel. North and his men made it back to Fenerive in 1709. Afterwards North sailed between Madagascar, Bourbon and the Comoros and some years later was killed by local tribesmen.

Halsey arrived at Fenerive in January 1708 and shared the booty. Some of the passengers of the *Essex* came afterwards with a small ship, the *Greyhound* with a license from the Governor of Madras to barter and recover some of their goods that had been captured. In the meanwhile came in a Scottish ship called *Neptune* (formerly called *Hannah*) mounting 26 guns and with 54 men commanded by Captain James Miller and with Samuel Burgess as first mate. The *Neptune* was to trade liquor for slaves then go to Batavia to dispose of the slaves and then to Malacca to recover the cargo of the *Speedwell* that was left on her return from China. All these people anchored in Fenerive were suddenly battered by a violent cyclone which destroyed all three pirates' ships, the *Buffalo,* the *Rising Eagle* and Thomas White's *Dorothy.* The *Neptune* survived but was obliged to cut away all her masts.

Having lost all their ships the pirates' thoughts were bent on seizing the *Neptune.* Burgess, who befriended Halsey, helped them in this design. He brought all her small masts and yards ashore. The pirates, having been requested to find the proper trees for masting asked for some extra hands. Captain Miller, suspecting no harm, came ashore with some of his crew. They were immediately seized and the long boat detained ashore. Miller was forced to send for the second mate and for the gunner. The mate, who was his brother, came but the gunner suspected foul play and refused. In the evening Burgess came on board the *Neptune* and advised the sixteen crew members remaining on board to surrender the ship they could neither defend nor sail. Two days later, the pirates manned the *Neptune*'s pinnace and seized the *Greyhound*. They took all the money they could find, put on board the Captain, second mate, boatswain and gunner of the *Neptune* and about 14 of her hands and some supplies and ordered her to sea. They detained the remainder of the *Neptune*'s crew.

While the *Neptune* was fitting, Halsey fell ill of fever and died. He was buried with great solemnity and ceremony which is immortalised by Johnson's epitaph: *"He was brave in his person, courteous to all his Prisoners, lived beloved, and died regretted by his own People. His Grave was made in a Garden of Water Melons, and fenced in with Pallisades to prevent his being rooted up by wild hogs, of which there are Plenty in those Parts."* After his death, David Williams was chosen to command the *Neptune*. He worked on the ship with great earnestness and made the Scot prisoners labour hard at fitting her up for a voyage. In 1709, she was refitted and ready to go to sea. However, another cyclone forced her ashore and she was wrecked before departing.

FLYING DRAGON[168] (1721)

Piracy had slackened in Madagascar until Woodes Rogers became governor of the Bahamas in 1718 and succeeded in taming piracy. This sent a new wave of pirates to Madagascar in 1719. It was Christopher Condent (sometimes named William "Billy One-Hand" Condon), originally from Plymouth who led the return to the eastern seas. Condent was second-in-command of a pirate sloop that fled New Providence in 1718. On the trip across the Atlantic, an Indian crewman who had been beaten and mistreated by the other pirates, threatened to blow up the ship's powder magazine. Condent leapt into the hold. The Indian discharged a piece at him, breaking his arm but Condent kept running at the Indian and shot him dead. In a barbaric display of relief at having been saved from an almost certain end, to use Johnson's words, "The Crew hack'd him to Pieces, and the Gunner ripping up his Belly, tore out his Heart, broiled and eat it."

Further into the voyage they captured a merchant ship called the *Duke of York*. The crew quarreled. The Captain and half of the men sailed away on the prize while the rest continued in the sloop and chose Condent as their Captain. At the Cape Verde Islands, Condent and his men took a Portuguese wine ship, an entire squadron of small boats and a Dutch warship. Condent kept the warship and renamed her *Flying Dragon*. The *Flying Dragon* marauded along the Brazilian coast for a while as Condent took more merchant ships including the *Wright* galley and the *Spelt* which he later sent away. Having heard that the Portuguese had imprisoned the crew of a pirate ship lost on the coast he tortured all the Portuguese who fell in his hands by "cutting off their Ears and Noses." He then sailed for the coast of Guinea capturing the *Indian Queen* commanded by Captain Hill and a Dutch ship. He next made for the East Indies taking an Ostend East Indiaman and a Dutch ship near the Cape. He reached Madagascar in June or July of 1719. While at Saint Marie, he picked up some of John Halsey's old crew into his own.

Condent went on to cruise the Indian coast and the Red Sea for a year or so. He first joined forces with other pirates at the Island of Anjouan and then sailed to the east coast of India. It was near Bombay in 1720 that Condent captured a large Turkish-owned ship sailing from India to Saudi Arabia packed with treasure and precious cargo to the value of more than £150,000 (worth about $375 million today) the richest prize in twenty years. Hoping to avoid the wrath of the already irate East India Company Condent ordered his men not to abuse the crew and passengers.

Condent returned to the island of Saint Marie, divided up the booty each man receiving about £2,000 ($5 million today), broke up their company and settled among the natives. They carried a petition to the Governor of Bourbon for a pardon, "tho' they paid the Master very generously." The Governor replied that he would take them into protection if they would destroy their ships which they agreed to. In 1721, after sinking the *Flying Dragon* in the harbor of St. Marie Condent and other members of his crew sailed to the island of Bourbon (Reunion). Twenty or more of the men settled there on the island. Condent, according to Johnson, went on to marry Governor Desforges Boucher's sister-in-law, traveled to France in 1723, settled down with his wife in St. Malo and became a wealthy merchant.

VICTORIEUX[169] (1723)

The *Victorieux* was John Taylor and Olivier le Vasseur's main prize. Le Vasseur was born in Calais around 1680. As a good womanizer he was given the name La Buse (*La bise*-The kiss). The first reference to him is with his brigantine the *Reine des Indes* in the company of Captains Benjamin Hornigold and Samuel Bellamy attacking both French and English ships near the Virgin Islands in 1716-17. After Woodes Rogers took over as Governor of the Bahamas in 1718, La Buse is among the pirates who surrendered and accepted the King's pardon. Shortly thereafter he departed the Bahamas and took up his old profession again. He is mentioned in 1719-20 near Wydah Road or Ouidah on the Bight of Benin along the Slave Coast where he plundered the area. In 1719, La Buse, seeing a ship anchored off Gambia Castle bore down upon it only to find that it was captained by Howell Davis another pirate. The two formed a partnership and sailing down the coast to Sierra Leone came up with Cocklyn a third pirate. The three crews took the fort and spent the next seven weeks refitting their ships. During this time they took an English slave ship, the *Bird Galley* commanded by Captain Jerrey Lecoole that came into the port. La Buse was given the galley in place of the brigantine he had previously commanded. He renamed her also *Reine des Indes* and mounted her with 24 guns. Having refitted, the pirates went to sea but an argument soon led the three to part company.

La Buse then set sail to the Indian Ocean and his first stop was the pirate haven of Saint Marie Island. The real chase could start. On the way to the Sea of Oman with little resistance he took the British Indiaman *Swanage* with a large booty. La Buse next sailed to the Maldives and the Chagos Archipelago. Heading for the Seychelles, danger loomed: he was chased by two warships. Fearful of being taken he fled. The two pursuers soon gave up. In fact they changed course to avoid an impending storm. This bad weather hit the *Reine des Indes* with full force. She nearly broke up. The sails that could not be hove down were torn and the rudder ruptured making her unsteerable. The *Reine des Indes* drifted with the current for 500 miles, lost half of her crew and eventually shipwrecked on Mayotte in the Comoros in July 1720. After trying to build a boat with the remaining timber, La Buse and half of his surviving crew were rescued by one of the long boats of the *Victory,* captained by John Taylor who arrived at Mayotte from Anjouan to recruit more men.

Taylor had originally served in the British Royal Navy, was disgraced during the reign of Queen Anne and had turned pirate. In 1719, quartermaster Taylor and his partner Captain Edward England had seized the *Peterborough* a 30-gun sloop on the west coast of

Africa, and renamed her *Victory*. They had just captured a large prize at Anjouan but at a high cost. On July 25, 1720 they surprised the 380-ton East Indiaman *Cassandra* mounting 36 guns and commanded by Captain Mackraw at anchor. Following an epic battle in which about 100 of the pirates were killed the *Cassandra* lowered her colors. England, as a gentleman and former Navy officer, spared Captain Mackraw against the wishes of the pirate crew. Mackraw seized the opportunity to flee in one of the boats. To permit an enemy of such a caliber to possibly come back and haunt them was an unforgivable crime to the crew. They enchained the compassionate Edward England and elected Taylor as Captain. Taylor renamed the *Cassandra* as the *Defense*.

After arriving at Anjouan with his men La Buse befriended John Taylor. Taylor proposed to take over the *Defense* and offered La Buse the command of the *Victory*. While unhappy about the loss of independence, La Buse had no choice and accepted. Together they sailed to the Indian Coasts and pillaged the inhabitants of the Laccadives Islands. Then the two ships sailed to Mauritius and arrived at Port Louis in February 1721 where they abandoned Captain England and three of his officers.

Two months later Taylor and La Buse sailed for Reunion Island. Arriving at Saint Denis in the early morning of April 20, 1721, they spotted a large Portuguese ship the *Virgem de Cabo* mounted with 72 cannons.[170] She had been there for three weeks repairing damage sustained during a storm en route from Goa. The majority of the 500-strong crew was enjoying the pleasures offered on land. Caught in a pincer movement by the two pirate ships which fired a few deadly broadsides the 100 men on board offered little resistance and quickly struck her colors. The booty was a pirate's dream—bars and coins of gold and silver, silk, diamonds and other precious stones, art, sacred vases and Goa's gold cross encrusted with rubies. Among the precious stones were 110 diamonds, some exceptionally large and beautiful, twenty large rubies, 250 emeralds and about twenty sapphires.

The *Virgem de Cabo* transported count Ericeira, Don Luis de Meneses, former Vice-Roy of the Portuguese East Indies with his fortune and the Archbishop of Goa. Both were on land. Taylor sent several boats to shore and asked for the Vice-Roy. Reluctantly he came aboard with two priests. He surrendered, handing over his sword whose handle was encrusted with diamonds. La Buse refused to accept it, telling him "Keep it; I bequeath it to you in memory of your unfortunate fate." Then he guided him to his cabin where an impro-

Figure 74. Captain Olivier le Vasseur, alias La Buse

vised music band played him a serenade before sending him back to shore.

After disarming the prisoners and taking on those who volunteered, entrusting the booty to the quartermaster and undertaking repairs, they sailed off. La Buse took the *Virgem de Cabo* in tow bound for Saint Marie Island. Taylor aboard the *Defense* seized the *Ville d'Ostende* (formerly *Greyhound*) of the Ostend East India Company on the way in the Bay of St. Paul. The booty was humbler but still worth the diversion. In Saint Marie, La Buse careened and refitted the *Virgem de Cabo*. They ripped up half a bridge to make her faster and reduced her armament to 60 guns. He renamed her *Victorieux*. At this point some of the newly enriched crew settled on Saint Marie.

Figure 75. Count Ericeira

Taylor and La Buse sailed on in early 1722, bound for India. In the open sea en route between Madagascar and Reunion, they took, pillaged and burned the *Princesse de Noailles,* a French East Indiaman used for local trade and picking up slaves in Madagascar. They rounded the south coast of Madagascar. While provisioning in St. Augustine Bay they discovered the letter left by Commodore Matthews with the specifics of the English squadron. Hurriedly, they changed course for the Mozambique coast. Taylor convinced the crews to aim for the land of gold (the kingdom of Monomotapa). They arrived on April 19, 1722 in front of the Dutch garrison at Fort Lagoa[171] which they took meeting little resistance as many in the garrison were ill. After setting in for two months, they sailed on June 30, 1722 after plundering the fort and taking with them a small hooker mounted with 12 cannons as prize and 16 of the Dutch settlers including the hydrographer Jacob de Bucquoy.[172]

Figure 76. Capture of Fort Lagoa by Taylor and La Buse (de Bucqoy, p. 28)

Emboldened, Taylor had a bigger design: take over the Portuguese settlement of Mozambique that held more riches than the capture of 100 vessels. La Buse was opposed to that venture arguing that it would be foolish to attack such a large port and settlement. During the night of 17-18 August 1722, La Buse and some of his officers tried to surreptitiously leave the company and sail back to the West

Indies. However, Taylor got wind of La Buse's intention. The next morning, a cannon shot was fired and the black flag displayed as a sign of distress. Taylor had La Buse and his accomplices judged and whipped by each of the crew and all their possessions seized to the benefit of the community. Nonetheless, a humiliated La Buse could keep the command of his vessel. Taylor was livid, treated everybody as cowards and then said: "Given that you don't seem willing to undertake manly enterprises anymore, let's go to the nearest land and everybody can then pursue their own fortunes." In unison the crew replied: "Let's go to Madagascar, and break up there."[173] The three ships crossed the Mozambique Channel and anchored at Bombetoke Bay (Mahajanga) on September 4, 1722.

The company broke up. Some stayed in Mahajanga with their wives and children under the protection of the King. Taylor took the *Defense* and prepared her for returning to the Caribbean where he intended to retire. The *Victorieux* was returned to La Buse who planned to cruise the Indian Ocean once again. Command of the hooker seized at Fort Lagoa was given to Captain Elck, a Scotsman, who planned to continue his pirate ways. The Dutch settlers captured in Lagoa were marooned, many died from illnesses and some survivors later reached Mozambique in a small barge. On November 4, 1722 the three ships set sail.

Taylor left for the Antilles seeking a pardon. He first put into Providence Island (off the coast of Nicaragua) but the Governor refused to pardon him. Next he sailed to Panama arriving in May 1723. Under duress, the Governor of Portebello pardoned him and his crew in exchange for the *Defense* and 121 barrels of silver coins. The pirates were just left with the diamonds and a little gold they could keep in their pockets, the only thing they earned from so many years of looting. Shortly thereafter, Taylor went to Jamaica where he still had a wife and four children. After spending all what he had he moved to Cuba, bought a plantation and a small boat and did some coastal trading between the islands.

La Buse returned to St. Marie. In mid-1723, he left for another run but only a few days out at sea the *Victorieux* was sighted by the English squadron of Commodore Matthews who gave chase. La Buse fled, panicked and committed a fatal navigation error. While sailing at full speed during the night the *Victorieux* hit a reef in the north of Madagascar close to Amber Cape when La Buse thought he was much further out. The ship was stuck, pounded by a heavy surf and broke up at low tide. There were about 125 survivors who built small barges out of the wood of the wreck. One day, while taking a nap around noontime they were attacked by the natives and many were murdered. Some survivors made it back to Bombetoke Bay (Mahajanga) with a barge; La Buse and others returned to St. Marie.

In late 1723, the news came that the French King offered a pardon to pirates provided they return their booty. On January 25, 1724 La Buse petitioned an amnesty for himself and 40 of his men to the Governor of Bourbon, Desforges Boucher. As a sign of goodwill he returned the religious artifacts captured on the *Virgem de Cabo* including the sacred vases to the Parish of St. Paul. His amnesty was granted the next day. However, La Buse's courier was mysteriously killed and he did not go to Bourbon. He renewed his request for amnesty in September and it was again granted. Many of his men moved to Bourbon in November 1724, but La Buse stayed behind in Madagascar, retiring in the Bay of Antongil.

In 1725, Governor Boucher dispatched a search party to Madagascar to entice the 40-odd pirates still there to come to Bourbon. He specifically asked to search for La Buse. The party met with La Buse at Antongil. Knowing the wrath against him of the prominent people of Bourbon, the men of the French East India Company because of the capture of the *Princesse de Noailles,* La Buse did not trust the Governor and refused to leave. With nothing better to do he became a pilot for ships wishing to navigate and anchor at Antongil. In 1730, when the *Méduse* with Captain L'Hermitte entered Antongil Bay, La Buse came forward and offered his services as pilot. L'Hermitte recognized him, put him in chains and took him to La Reunion where he was tried and hanged at Saint Denis on July 17, 1730.[174] When about to be hanged he threw a piece of parchment to the gathering, shouting *"find my treasure who can."* To this day, the treasure has not been found.[175]

REFERENCES

[138] Mack, J., op. cit., p. 13, citing Ferrand, G. (with revisions by Vérin), "Madagascar," in Bosworth, C.E., van Donzel, E., Lewis, B., and Pellat, Ch. (eds.), The Encyclopedia of Islam, Leiden: Brill, 1986, Vol. 5, p. 940.

[139] There is a vast array of books, articles, and web sites devoted to piracy and their presence in Madagascar. Some that were used for this background section include: Captain Johnson, Ch., "A General History of the Pyrates," London: Vol. I: T. Warner, and Vol. II: T. Woodward, 1724; de Bucquoy, J., "Zestien Jaarige Reize naar de Indiën gedaan door Jacob de Bucquoy, vol Aanmerke-

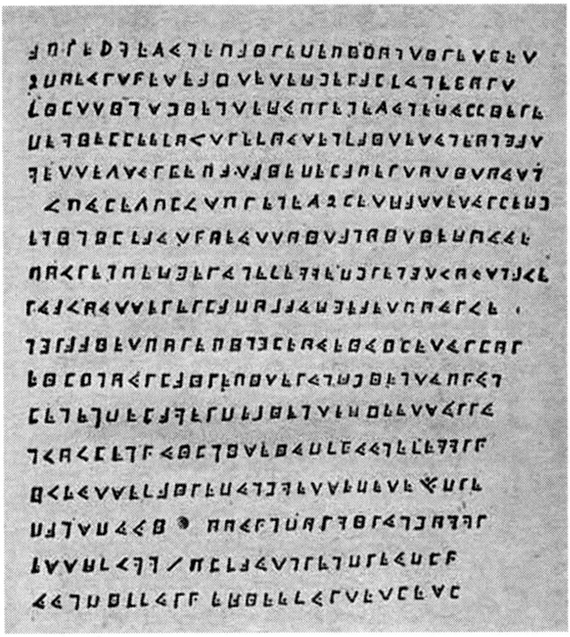

Figure 77. La Buse's cryptogram

lyke Ontmoetingen," Haarlem: Jan Bosch, 1758 (second ed.); Brockway, T., "The pirates in Madagascar," Antananarivo Annual, 1886, p. 250; Grandidier et al., op. cit., t. III, pp. 450-456; Deschamps, H., "Les Pirates à Madagascar aux XVIIème et XVIIIème Siècles," Paris: Berger-Levrault, 1949 (and second edition, 1972); Cordingly, D., "Life among the Pirates: The Romance and the Reality," London: Little Brown & Co., 1995; Bastions Pirates, Do or Die, Ed. Aden, No. 8 (2001), reproduced at http://www.eco-action.org/dod/n08/pirate.html and a French translation at http://mathieu.saura.free.fr/site/texts/text%20bastions%20pirates.htm; Chris Rule, "Pirate strongholds and hideouts," at http://www.piratesinfo.com/detail/detail.php?article_id=70.

[140] Hood, J., op. cit., p. 160.

[141] Sakolsky and Koehnline (eds.), "Gone to Croatan: The Origins of North American Dropout Culture," New York/Edinburgh: Autonomedia/AK Press, 1993, p. 107.

[142] A good description of the rules aboard a pirate ship is provided in de Bucquoy, op. cit., pp. 65-71. See also Snelgrave, "A new Account of some parts of Guinea and the Slave-trade," 1734.

[143] See, *inter alia,* Bolster, W. J., "Black Jacks: African American Seamen in the Age of Sail," Harvard University Press, 1997, pp. 12-14.

[144] Burg, B. R., "Sodomy and the Pirate's Tradition: English Sea Rovers in the Seventeenth Century Caribbean," New York, 1984, claims that the majority of

the pirates were homosexuals. There is no solid proof to support this theory. However, certain buccaneers lived in a type of homosexual union called "matelotage" (from the French "matelot" or sailor, believed to be the origin of the term "mate"), sharing their belongings, with the survivor inheriting his companion's share in case of death. Even when women joined the buccaneers, the practice continued, with the sailor sharing his wife with his partner.

[145] Captain de la Merveille, « Mémoire manuscrit » addressed to Minister de Pontchartrain, 1712, kept at the French Archives Coloniales, Correspondance générale, Madagascar, carton 1, pièce 31.

[146] Downing, Cl., "A Compendious History of the Indian Wars," 1737, pp. 52, 65, 80, and 235.

[147] Sources: Dampier, W., "A New Voyage Round the World," London, 1697; Ovington, J., "A voyage to Suratt in 1689;" Letter by Captain Willian Freke of the *Anne,* December 8, 1689, British Library, India Office and records, No. 5690, pp. 93-98; Wikipedia, Charles Swan, at http://en.wikipedia.org/wiki/Charles_Swan; Kingston, W.H.G. and Frith, H., Notable Voyagers, Chapter XXII, Dampier's voyages, at http://www.athelstane.co.uk/kingston/voyagers/vyage22.htm; A Buccaneer's Atlas, Basil Ringrose, at http://content.cdlib.org/xtf/view?docId=ft7z09p18j&chunk.id=d0e1879&toc.depth=1&toc.id=d0e1720&brand=eschol; Captain Swan at http://www.burleygames.com/Captain swan.htm; William Dampier, NNDB, at http://www.nndb.com/people/943/000096655/.

[148] While in the Galapagos, Dampier was the first to provide us with an accurate description of the islands and their fauna and flora.

[149] Also called *Johor and Rebecke.* Sources: Report by Jan Coin, Captain of the yacht *Tamboer,* 1698, in "Tweede deel der brieven en papieren van de Kaap [de Goede Hoop] overgekomen," ff. 820-825, Amsterdam: Rijksarchief, Koloniaal Archief, 4020; Drury, R., "Madagascar: or Robert Drury's Journal during fifteen years captivity on that island," London: W. Meadows, 1729, pp. 105-116; Molet-Sauvaget, A., "Un Européen, roi 'légitime' de Fort Dauphin au XVIIIe siècle: le pirate Abraham Samuel," Etudes Océan Indien, 1997, 23/24: 211-221; Cindy Vallar, "Black Pirates," in Pirates and Privateers, The History of Maritime Piracy," at http://www.cindyvallar.com/blackpirates.html; Privateers Dragons of the Caribbean, at http://www.privateerdragons.com/pirates_famous3.html#H.

[150] These included two English East Indiamen that were ransomed, and the *Rouparelle.*

[151] Sources: Captain Johnson, op. cit., t.II, pp. 65-80; Zacks, R., "The Pirate Hunter, The True Story of Captain Kidd," New York, 2002; Clifford, B., "Return to Treasure Island and the Search for Captain Kidd," New York: Harper Collins, 2003; Captain William Kidd, USS KIDD Veterans Memorial, Louisiana Naval War Memorial Commission, 2006, at http://www.usskidd.com/willkidd.html; Chris Rule, "Piratical History of Madagascar," at http://www.piratesinfo.com/detail/detail.php?article_id=70; Paul Hawkins, "The

Ultimate Captain William Kidd website" at http://www.Captain kidd.pwp.blueyonder.co.uk/.

[152] The *Rouparelle* had already been seized in 1696 by John Hoar on the *John and Rebecca.*

[153] Another theory, advanced by Paul Hawkins, op. cit., is that Kidd and Culliford colluded. Kidd would have cached a large part of his booty on an island hideaway on the way to St. Marie. To quote Hawkins: "When Kidd arrived at St. Marie, he was delighted to see Culliford's ship the *Mocha Frigate* at anchor in the harbor. Kidd, now aboard the *Quedagh Merchant,* no longer needed his leaking fighting ship and offered Culliford the *Adventure Galley,* and the members of his crew who demanded to stay in the Indian Ocean, in return for replacement crew, goods and money. Culliford, whose ship *Mocha Frigate* was severely damaged in a skirmish with the British ship *Dorrill* in the Malacca Straits (her main mast was sheared by a salvo from the *Dorrill*), readily agreed and set about stripping the *Adventure Galley* to make her lighter to haul on to the careening beach. Kidd and Culliford soon forgot about past indiscretions, and enjoyed each others company in the taverns of St. Marie whilst Culliford waited for the *Adventure Galley* to be made sea-worthy. After Culliford had the *Adventure Galley* careened (which would have taken him about six weeks), he and his newly acquired crew set sail in the *Adventure Galley* (which would have been re-named) to resume their acts of piracy in the Indian Ocean. Kidd then stripped the *November* of everything useful, sold off any goods of value from her hold and torched her in the harbor, along with the badly damaged *Mocha.*"

[154] Sources: Privateers Dragons of the Caribbean, at http://www.privateerdragons.com/pirates_famous1.html#C; Rob Ossian, "Pirate's Cove," at http://www.thepirateking.com/bios/chivers_dirk.htm.

[155] Sources: Durup, J., "Short seafaring adventures and conflicts in the Indian Ocean 1405-1811," 2004, at http://perso.orange.fr/henri.maurel/seafaring%201.htm; Rob Ossian, "Pirate's Cove," at http://www.thepirateking.com/bios/stout_ralph.htm.

[156] Certain reports indicate that the *Mocha* was then renamed the *Defence.*

[157] A number of reports state that Culliford changed the ship's name to the *Resolution.*

[158] Sources: Captain Johnson, op. cit., pp. 373-413; Pirate Encyclopedia, Nathaniel North at http://ageofpirates.com/article.php?Nathaniel_North.

[159] Williams later participated to the capture of the *Speaker,* and then sailed with the *Prosperous* and the *Charles.*

[160] A former Buccaneer, Burgess also helped in stealing William Kidd's *Blessed William* in February 1690 in Nevis. Afterwards, he worked for Frederick Phillips doing trade runs to Madagascar, where he sold the pirates supplies and guns for gold and slaves. He later came back to Madagascar with the *Neptune.*

[161] Sources: Captain Johnson, Ch., op. cit, vol. II, pp. 240-250; Privateers Dragons of the Caribbean, at http://www.privateerdragons.com/pirates_famous3.html#C.

[162] Also called Murder Island, because of the assassination there of two officers of the British Navy ship *Barracouta* on May 22, 1824. These two officers were interred with military honors on a nearby islet, Nosy Andrambala, also called Grave Island. Johnson locates the reef "40 miles north of St. Augustine Bay." However, the first small island North of St. Augustine Bay is Nosy Hao, about 90 nautical miles north of St. Augustine.

[163] For a description of that wreckage and archaeological investigation, see Lizé, P., "Piracy in the Indian Ocean, Mauritius and the Pirate Ship *Speaker*," in Skowronek, R.K. and Ewen, C.R. (Editors), "X Marks the Sport, The Archaeology of Piracy," University press of Florida, Gainsville, 2006.

[164] Sources: Captain Johnson, op. cit., vol. II, pp. 110-118; Durup, J., "A short seafaring adventures and conflicts in the Indian Ocean 1405-1811," 2004, at http://perso.orange.fr/henri.maurel/seafaring%201.htm; Wikipedia, "John Halsey" at http://en.wikipedia.org/wiki/John_Halsey.

[165] See above, under the English East India Company.

[166] As mentioned above, White, with George Booth, John Bowen and Nathaniel North, had captured the *Speaker* (originally a French warship, captured and refitted as an English merchantman) in 1700 in Majunga. After returning from a run in the Red Sea and the coast of Malabar, and spending some time in Madagascar, in 1703 White teamed up with some of the surviving crew of the *Degrave*. In the Red Sea they captured first a large Moorish vessel, the *Malabar,* and then the *Dorothy,* a 225 ton ship that was in service of the HEIC from 1687 to 1698 and had undertaken four voyages to India. In 1703, the *Dorothy* was sailing under the command of the English Captain Penruddock, but with a Moorish crew. When the pirates came alongside her, the men of the *Dorothy* offered no resistance. On a vote, White's crew gave Captain Penruddock (from whom they took a considerable quantity of money) the *Malabar,* while keeping the *Dorothy* for their own use.

[167] The *Rising Eagle* was a 600-ton east Indiaman mounting 16 guns, belonging to the New Company and put in service in 1700. She sailed to Balasore in Eastern India on her maiden voyage.

[168] Sources: Captain Johnson, op. cit., pp. 139-143; Chris Rule, "Pirate strongholds and hideouts," at http://www.piratesinfo.com/detail/detail.php?article_id=70; Wikipedia, "Christopher Condent" at http://en.wikipedia.org/wiki/Christopher_Condent.

[169] Sources: de Bucquoy, op. cit., pp. 27-112 ; Bibique, « Sur la piste des frères de la côte, » Paris : Orphie, 1988; Lougnon, A., « Sous le signe de la Tortue, voyages anciens à l'île Bourbon (1611-1725), Paris : Azalées Editions, 1992; Viala, G., « La Buse, un pirate dans l'Océan Indien, » Ed. Du paille-en-queue noir, 1997; Davis Stapleton, "Pirate Roster—Olivier La Bouche », 2001, at http://piratesold.buccaneersoft.com/roster/olivier_la_bouche.html;   s.n.,

« Olivier Le Vasseur, dit La Buse, le Pirate, » at http://dossiers.clicanoo.com/article.php?id_article=97577&id_mot=.

[170] Originally a Dutch man-of-war named *Gelderland,* she was sold to the King of Portugal.

[171] At Maputo Bay, formerly called Baia de Lourenço Marques, an inlet on the coast of Mozambique. In 1720 the Dutch East India Company built a fort and factory called Lijdzaamheid (Lydsaamheid) there, since April 1721 governed by an Opperhoofd (Chief factor) under authority of the Dutch Cape Colony, interrupted by Taylor's occupation from April through June, 1722; in December 1730 the settlement was abandoned.

[172] Who wrote a full account of these events (op. cit.), first published in 1744.

[173] Cited by de Bucqoy, op. cit., pp. 44-45.

[174] The traditional account was that La Buse was tried and hanged in St. Paul on July 7, 1730. This is not correct, as several reports clearly indicate that the trial and hanging took place in St. Denis on July 17, 1730, including: a letter from Mr. Dumas, Governor of Bourbon, to Minister de Maurepas, dated December 29, 1730, kept in the Centre des Archives d'Outre Mer, Aix en Provence, Correspondance générale de Bourbon, t. V, 1727-1731; a letter from the Council in Bourbon to the East India Company dated December 20, 1730, reproduced in Lougnon, A., "Correspondence du Conseil Supérieur de Bourbon et de la Compagnie des Indes, 22 janvier 1724—30 décembre 1731," St. Denis (Réunion), 1934, p. 131; and Guët, I., "Les Origines de l'île Bourbon et de la colonisation française à Madagascar," 1886, p. 218; see also "La Buse pendu à Saint-Denis," Le Quotidien de la Réunion, July 3, 2007, p. 8.

[175] There is some confusion in the 18th century literature between Condent and La Buse. Johnson, in particular, does not mention La Buse. Startlingly, the stories of both accepting Woodes Rogers' pardon and then fleeing the Bahamas in 1718 to continue their pirate ways are identical. Johnson relates that Condent captured the *Indian Queen* at the coast of Guinea in early 1719. At the same time and along the same coast, La Buse is reported to have captured a galley renamed *Reine des Indes (Indian Queen* in French). Johnson also mentions that, after coming to Madagascar, Condent "took, in company of two other Pyrates he met at St Mary's, the *Cassandra* East-India Man," then "returning, touch'd at the Isle of Mascarenas, where he met with a Portuguese Ship of 70 guns, with the Vice-Roy of Goa, on board," and next "carried the Prize to the Coast of Zanguebar, where there was a Dutch fortification, which they took and plunder'd, razed the Fort, and carried off several Men who enter'd voluntarily," all of which are attributed to La Buse in other narratives.

6

British Royal Navy

Beyond the four squadrons to suppress piracy in the early 18th century, the British Royal Navy presence around Madagascar was sparse. This changed after the French Revolution broke out. Initially Britain did not get involved. However, after the belligerent French revolutionary government had executed the King it declared war on February 1, 1793 starting 22 years of hostilities that became know as the Great European Conflict (subdivided into the French Revolutionary War 1793-1802 and the Napoleonic Wars 1803-1815). Britain used these wars to make significant colonial gains including in the Indian Ocean and to strengthen its hold in India. The British Navy occupied the Seychelles in 1794 and greatly increased its activity in the Indian Ocean through patrols and naval engagements notably around the French possessions in Bourbon and Mauritius. Also, as a protection against the French navy and privateers towards the end of the 18th century it became customary for British East Indiamen to sail in convoy with a Royal Navy escort. Bourbon surrendered to the British in July 1810 and Mauritius was captured on December 3, 1810 under Commodore Josias Rowley. British possession of Mauritius was confirmed four years later by the Treaty of Paris (1814) resulting in a permanent presence of the Royal Navy in the lower Indian Ocean. Bourbon was returned to the French government.

SIBYL/GARLAND[176] (1798)

The *HMS Sibyl* was a 6th rate Royal Navy ship mounting 28 guns, built in 1779 at Bucklers Hard. She first sailed with Admiral Sir William Pasley from Portsmouth to St. John's, Newfoundland and then to Lisbon. In July 1795, she was renamed *Garland* and sailed under Captain John Erskine Douglas in the North Sea. When he removed into Boston at the beginning of 1798, her command befell to Captain James Athol Wood. Captain Wood was sent by Rear Admiral Sir Hugh Christian to cruise off Mauritius and Bourbon with a small squadron. There he received intelligence that two large French frigates which had been causing great depredations in the Indian Ocean were sailing towards Madagascar.

The squadron went in pursuit and on July 26, 1798 Captain Wood discovered a large vessel anchored near the former French settlement of Fort Dauphin. Because the rest of the ships were to windward and unable to work against the current *Garland* went in to examine her. When she was within about a mile *Garland* struck on a pointed rock about 15 feet below the surface 5 miles northeast from the Islet of St. Luce. Water immediately poured in through the midship ports on the main deck and the hawse holes and all Captain Wood could do was get the whole crew into the boats and save the stores and rigging.

The enemy ship was not a frigate but a large merchant ship pierced for 24 guns with a complement of 150 men. They had run her ashore when *Garland* approached but as soon as they saw she was in trouble they tried to regain their vessel. *Garland*'s boats, being to windward, reached and secured her first. Captain Wood was successful in befriending the natives and they delivered most of the Frenchmen to him as prisoners. Over the next four months he built one vessel of 15 tons and had made good progress on another when the *Star* sloop of war arrived at St. Luce. The French prisoners of war were taken to the Cape of Good Hope in her. The *Garland*'s officers and men returned in their prize. During his time in Madagascar, Captain Wood surveyed the coast from Fort Dauphin to St. Luce and discovered an anchorage within the reef capable of holding a fleet of battle ships.

BLENHEIM AND JAVA (1807) AND HARRIER[177] (1809)

HMS Blenheim, a three-decker armed with 90 guns was built in 1761 at Woolwich and put out of commission at Chatham in 1799. In

1801, after being inspected by the Navy Board *Blenheim* was ordered to be cut down to a 74-gun ship. She was employed as a guard ship at Portsmouth in 1802 and then sailed for the West Indies late that year capturing some French privateers near Martinique in 1803. In 1805, by then an old worn-out ship she became the flagship of Vice Admiral Sir Thomas Troubridge. Born in London about 1758, Sir Thomas Troubridge had a brilliant naval career. In March 1801, he became a lord of the admiralty was promoted to the rank of Rear-Admiral on April 23, 1804 and retired from the admiralty in May 1804. In April 1805 he was appointed to the chief command in East Indian seas to the eastward of Point de Galle and went out for the East Indies with his flag in the *Blenheim* under the command of Captain Austen Bissell.

Shortly after passing Madagascar and having with him a convoy of ten Indiamen, Troubridge fell in with the French admiral Charles de Durand-Linois, Commander of the French Naval forces in the Indian Ocean in the *Marengo,* with two large frigates in company, le *Bélier* and la *Belle-Poule.* Linois probably mistaking the *Blenheim* for an Indiaman approached with a view to seize a rich prize. Finding out his mistake and notwithstanding the disparity of force he hauled his wind and made off. Even if the *Blenheim* was a ship to chase with Troubridge would not have felt justified to leave the convoy and also knew that the chase would be useless. He pursued his voyage and joined in Penang Sir Edward Pellew till then commander-in-chief in East India and China.

Pellew was strongly convinced of the inadvisability of dividing the station as the exigencies of war might make prompt action under one commander essential to success. Troubridge maintained that they had no power by any agreement between themselves to alter the disposition of the admiralty. So Pellew referred the matter to them. The result was an order to Pellew to resume the command of the whole station and to Troubridge to take the chief command at the Cape of Good Hope. Miffed, Troubridge wanted to take up his new station as fast as possible. There was no quarrel but because of the blunder of the admiralty the relations between them were not altogether friendly.

In the meantime, the *Blenheim* was ashore in the Straits of Malacca and had sustained so much damage that in the opinion of many of her officers she was no longer seaworthy. At Troubridge's insistence the *Blenheim* sailed at the end of 1806, after running aground at the entrance to the Straits of Penang, under jury masts from Pulo Penang to Madras to refit. There she was found to be hogged and totally unfit for sea. In fact, the pumps were barely able to cope with the water ingress when she was at anchor. Her Captain, Bissell, stated that there would be great danger in attempting to take her to the Cape. Troubridge was unwilling to remain on Pellew's station longer than necessary. He insisted on sailing at once in the *Blenheim* and his confidence was buttressed by the fact that many passengers from Madras embarked in her. *Blenheim* left Madras on January 12, 1807 with *Java,* an old Dutch prize frigate under the command of Captain George Pigot, and the *Harrier* brig which in 1806 was under the command of Troubridge's son, Edward Thomas Troubridge in the East Indies.

The *Java,* a 5th rate Royal Navy ship mounting 32 guns was originally named *Maria Reygersbergen.* She was a Dutch prize frigate taken into the British Royal Navy after being captured in 1806 by *Caroline* under the command of Captain Peter Rainier. *Caroline* was off Batavia (Djakarta) on October 18, 1806 and in the morning captured a small brig from Bantam. From her Captain Rainier learned that a Dutch frigate, the *Phoenix,* was under repair at Omust and resolved to bring her out. Between Middleby and Amsterdam Island he discovered two brigs at anchor and captured one the *Zeerop* with 16 guns. The other made her escape into Batavia where she joined the *Maria Reygersbergen* frigate, the *William* 20-gun sloop, the 18-gun *Patriot* and the 14-gun *Zeeplong. Caroline* ran for the *Maria Reygersbergen* and opened fire when within half-pistol-shot. After about half an hour the Dutch frigate hauled down her colors. She was armed with 36 guns, 18-pounders on the main deck and carried a crew of 270 men. *Caroline* was 57 below complement with men away in prizes or sick in hospital. Because the action took place in four fathoms amid dangerous shoals it was not possible to capture the other vessels but they and the *Phoenix* were seen to run themselves on shore.

On February 5, 1807 near the southeast end of Madagascar, the three ships got into a cyclone from which the *Harrier* alone emerged. When last sighted both the *Blenheim* and *Java* had hoisted signals of distress. But the *Harrier* herself was in great danger and could do nothing. She lost sight of them in a violent squall with the *Blenheim* appearing to be settling in the water. It is possible that *Java* ran foul of the sinking *Blenheim* while trying to rescue Sir Thomas Troubridge. Both ships foundered with all hands.

The joint crews of the two ships amounted to at least 1,000 people. One of the victims was the 46-year old James Morrison. He was one of the mutineers of the *Bounty* and the *Bounty*'s own intellectual protestor. After the failed attempt to build a colony on Tubuai, Morrison lived in Tahiti for a while. He returned to England, where he was tried and found guilty by the court but with the highest recommendation of mercy. He immediately returned to active duty as a Warrant Officer and served during the heroic period of St. Vincent, the Nile, Copenhagen and Trafalgar. His last ship was the *Blenheim* on which he had served as a young gunner prior to his *Bounty* experience.

When the news reached the East Indies, Pellew sent Troubridge's son then in command of the *Greyhound,* to make inquiries as to the fate of the ships. Edward Troubridge cruised to Mauritius, Madagascar and the Cape in a desperate search for his father. The French governor of Mauritius assisted him as well as he could and sent an account of pieces of wreck which were cast ashore in different places. But nothing could be identified as belonging to either of the missing ships and nothing gave any positive information as to their fate. A heartbroken Edward Troubridge returned to India.

Meanwhile, *Harrier* made it to the Cape of Good Hope with great difficulty. She was refitted. During March 1809 under the command of Captain John James Ridge while sailing back to India she foundered in the same vicinity of Madagascar. There were no survivors.

STAUNCH[178] (1811)

HMS Staunch was a 182-ton Royal Navy Gun-brig armed with 14 cannons built in 1804 by Tanner at Dartmouth. Her story is intimately linked with the capture of Mauritius. After sailing the English Channel in 1805, *Staunch* sailed for Cape of Good Hope on August 30, 1806 under the command of Lieutenant Benjamin Street. When it was learned that the Spanish colonists had recaptured Buenos Aires vessels were sent from the Cape with transports to Argentina. *Staunch* took part in the attack on Montevideo in January 1807. About 800 seamen and Royal Marines were landed to act with the troops. The Spanish put up such a strong resistance that the squadron had only two days powder and shot left when a breach was made on February 3, 1807 and the town and citadel were taken by storm.

Staunch then returned to the Cape and proceeded to the Indian Ocean. Shortly after the surrender of Reunion in July 1810 eyes were set on the capture of Mauritius. The British Navy ships *Nereide, Sirius* and *Staunch* sailed for Mauritius in late July. They first attacked the Isle de la Passe, a small island off Grand Port on the southeast side of Mauritius. On August 13, the French garrison surrendered and the British occupied the batteries on the islet. On August 17, they landed at Canaille de Bois near Grand Port. After a march of six miles, attended by three of the *Nereide*'s and *Staunch*'s boats with guns mounted covering the road, they attacked and carried a fort at Point du Diable which commanded the northeast passage into Grand Port. At nightfall they returned to their ships but landed again on three successive days to destroy the guns in the fort and the signal house at Grande Rivière. Meanwhile, the *Staunch* had left them and proceeded to Port Louis on the other side of the island.

On August 20, the entire situation which until then appeared so favorable for a speedy conquest of the island was suddenly changed with the unexpected approach of a squadron of five French Navy ships. The French ships managed to anchor off Grand Port and epic battles ensued. Reinforcements were called in including the *Staunch*. On August 23rd the British fleet attacked down the channel into Grand Port. It became a rout with their ships running aground or being blown up by the superior force of the French. On August 28 the British surrendered.[179] The *Staunch* and the other ships that could be saved sailed back to Reunion.

Desiring to take full advantage of their success at Grand Port the French formed a squadron to attack Reunion. On the morning of September 12, 1810 *Staunch* and the other British ships weighed anchor from the Bay of St. Paul in Reunion in order to attack the approaching French frigates. The French fled and *Staunch* helped recapture the *Africaine* that was taken by the French a few days earlier. Another battle with two French ships took place close to Reunion on September 18. In October 1810, *Staunch,* now under Lieutenant Henry Craig, and two other British ships were detached to convoy troops from Bourbon to Rodrigues Island which was previously occupied as a base for the British blockading squadrons off Reunion and Mauritius. The convoy arrived on November 12. After being joined by more troops from Bengal on November 29 the whole fleet of nearly 70 sails anchored in Grand Bay about 12 miles to windward of Port Louis and marines and seamen disembarked. On December 2, 1810 the French Governor proposed terms of capitulation and these were ratified the following day.

After the capture of Mauritius, *Staunch* continued to sail under Lieutenant Henry Craig. During June 1811 she was wrecked off Madagascar with all hands lost.

REFERENCES

[176] Source: Grocott, T., "Shipwrecks of the Revolutionary and Napoleonic Eras," London: Caxton, 2002, p. 59; Michael Phillips' ships of the old Navy, at http://www.ageofnelson.org/MichaelPhillips/info.php?ref=1004.

[177] Sources: Sailing ships of the Royal Navy, http://www.cronab.demon.co.uk/; Michael Phillips' ships of the old Navy, at http://www.ageofnelson.org/MichaelPhillips/info.php?ref=0347; Sir Thomas Troubridge, at http://www.aboutnelson.co.uk/13troubridge.htm; Grocott, op. cit., pp 233-235 and p. 277.

[178] Sources: James, W., "The Naval History of Great Britain from the Declaration of War by France in February 1793 to the Ascension of George the IV in January 1820," London: Harding, Lepard & Co., 1826, Vol. V, p. 474; Norie, J. W. (compiler), "The Naval Gazetteer, Biographer and Chronologist, Containing a history of great Wars from their commencement in 1793 to their conclusion in 1801; and from recommencement in 1803 to their final conclusion in 1815," London: J.W. Norie & Co., 1827, pp. 67-68 and 478; Gosset, W. P., "Lost Ships of the Royal Navy, 1793-1900," London: Mansell, 1986, pp. 80 and 148; Sailing ships of the Royal Navy, http://www.cronab.demon.co.uk/; Michael Phillips' ships of the old Navy, at http://www.ageofnelson.org/MichaelPhillips/info.php?ref=2125.

[179] This battle at Grand Port became Napoleon's sole naval victory over the British. It is commemorated on the Arc de Triomphe at the Place de l'Etoile in Paris.

7

French Navy

HMS SERAPIS[180] (1781)

HMS Serapis, a British frigate launched by the Royal Navy in 1779 was designed as a 5th rate vessel armed with 44 guns (twenty 18-pounders, twenty 9-pounders and four 6-pounders). In September 1779, commanded by Captain Richard Pearson she and the *Countess of Scarborough* 22 guns escorted 40 British merchantmen returning from the Baltics. On September 23, Captain John Paul Jones aboard the American warship USS *Bonhomme Richard*[181] sighted the convoy in the North Sea off Flamborough Head, England and signaled the attack. Most of the merchantmen ran in shore and anchored under the guns of the *Countess of Scarborough.* Around 6 pm, the *Bonhomme Richard* engaged the *Serapis.* The battle raged on for more than three hours as the crew of *Bonhomme Richard* tenaciously fought the *Serapis* raking her deck with gunfire.

At first, a British victory seemed inevitable as the more heavily armed *Serapis* used its superior firepower to rake the *Bonhomme Richard* with devastating effect. When Captain Pearson called out to Jones asking if he struck his colors Jones replied with those immortal words; "Struck? I have not yet begun to fight!" The Americans continued the fight with light guns on the spar deck and eventually pulled alongside and lashed the two ships together. An attempt by the Americans to board *Serapis* was repulsed as was an attempt by the British to board the *Bonhomme Richard*. Finally an American party under the command of Nathaniel Fanning seized control of the enemy tops and used this position to clear the weather deck below with grenades, mortars and gunfire. Finally, a hand-grenade was dropped from the main-yard of the *Bonhomme Richard* down a hatchway in the *Serapis,* causing a violent explosion on the lower deck and the *Serapis* struck its colors.

The lives of nearly half of the American and British crews were lost. The *Bonhomme Richard,* shattered, on fire and leaking badly defied all efforts to save her and sank two days later on September 25, 1779. John Paul Jones transferred his command to the captured *Serapis*. She was rigged with jury masts and ten days later reached Texel in the Netherlands for repairs. This pivotal engagement, no less in home waters stung the British admiralty and gave the American cause encouragement during the dark days of the Revolution.

As the *Bonhomme Richard* was loaned to Benjamin Franklin to aid the American Revolution, orders were given to Captain Cottineau de Kerloguen to take command of the *Serapis* in Texel and bring her back to France.[182] The *Serapis* was refitted for the French navy and sent to the Indian Ocean in 1781 under Lieutenant de Vaisseau Roche. After an uneventful sailing down the Atlantic Ocean, she put in at False Bay at the Cape settlement on May 19, 1781 to secure supplies and disembark 40 ill sailors. She left False Bay on June 14 and anchored in the harbor of Ambudifutatra at St. Marie Island on July 31.

At two in the afternoon on the day of the arrival the first mate, Mr. Lheritier, descended in the backhold on the starboard side to thin eau-de-vie. Against strict orders he demanded that the lights be taken out of the lanterns to see more clearly. A spark, or rather a small portion of the burning wick fell into the funnel that he was using and the spirit immediately caught fire. A thick and burning smoke rapidly spread throughout the hold suffocating the crew attempting to extinguish the fire.

Captain Roche, who was ashore, was alerted by a cannon shot and immediately went back on board. He found the hold ablaze. He tried to have the steerage cut to allow water into the hold but this proved impossible. He then ordered to open the taps to sink the ship but this did not succeed. He decided to beach the ship but seeward winds prevented cast off. After all attempts at saving the *Serapis* failed at 6 pm Captain Roche ordered all the men aboard the boats that had already been swung out. The evacuation was orderly and when all men where on the boats the Captain left her. The *Serapis* exploded 45 minutes

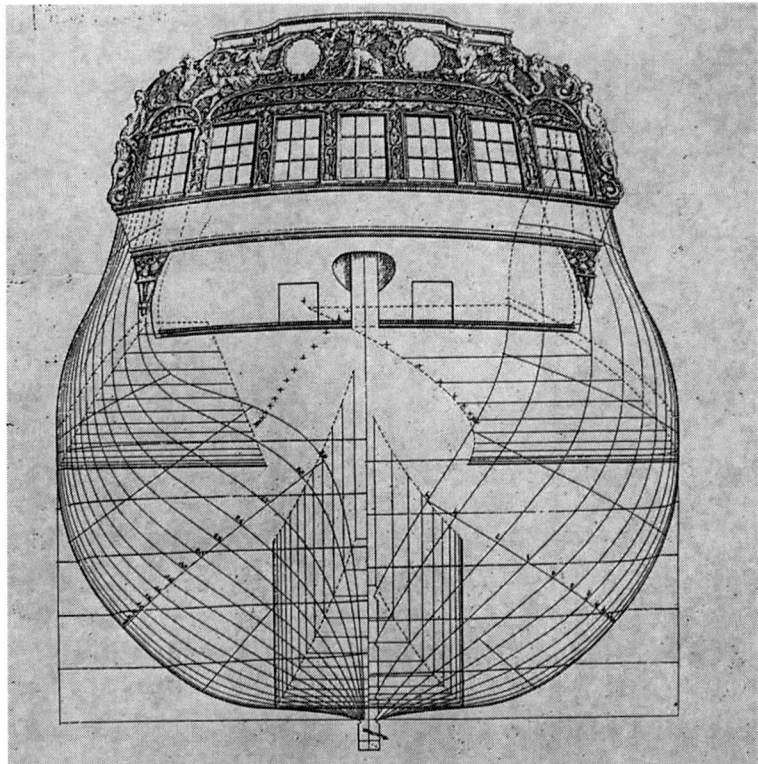

Figure 78. Etch of the *Serapis'* poop

later and immediately sank. Only two men died; one bedridden with fevers and the Master Gunner who drowned after having filled his pockets with 2,000 piasters which he had stolen for his own use. The first mate was arrested and the crew transferred shortly thereafter to Mauritius.

In 1999 a crew of American, French and Malagasy archaeologists and historians lead by Dick Swete located the remains of the vessel. In May 2004, a dive team returned to Ste. Marie to research the vessel remains under the direction of Michael Tuttle. They salvaged a copper box which has been restored by this author.

CHEVRETTE[183] (1830)

The Navy vessel *Chevrette* spent the bulk of her career on discovery and charting expeditions. In 1819, the famous explorer Jules Dumont d'Urville sailed on her to survey the Black and Aegean Seas.

Figure 79. Copper box recovered from the *Serapis*

She later sailed in Asia, and in 1827 transported 300 slaves from New Guinea to Bourbon.[184] In 1827-1828, under the command of Captain Théodore Fabré, she undertook a mission to survey the coasts of India for the French Navy. Thence she joined the naval squadron based in Bourbon Island.

The *Chevrette* became the first naval victim of the troubled relations between France and Madagascar throughout the reign of Queen Ranavalona I (1828-61). Humiliated by the surrender to King Radama I of the French settlements set up by Sylvain Roux in the 1820s in Tamatave, Foulpointe, Tintingue (Manompana) and Fort Dauphin, the decision was taken in Paris on January 28, 1829 to send an expedition to Madagascar under the command of the Capitaine de vaisseau Jean-Marie Goubeyre. In July 1829 five Navy ships showed up in front of Tamatave. Several battles took place in the ensuing months in Tamatave, Foulpointe, Tintingue and Antsiraka (Pointe à Larrée) with varying successes for both parties.

In the night of March 20, 1830, the *Chevrette,* under the command of Lieutenant de vaisseau Hyppolite Depanis, left the harbor of Tintingue in heavy weather and was thrown against the reef at the entrance of the channel. The *Chevrette* immediately burst open and was lost. An ensuing court martial acquitted Captain Depanis, recognizing that the events had been beyond his control.

COLIBRI[185] (1843)

On February 25, 1843 at five-thirty in the morning, three French navy vessels, the corvette *Berceau, le Voltigeur* and *le Colibri* set sail from Anorotsangana on Madagascar's west coast. Weather was nice with a light northwesterly wind. They kept the luff to round the Radama islands. Early in the afternoon however, the winds freshened and the sea became heavy. Captain Orcel of the *Colibri* had the square topsail and the main sail reefed. He was sounding while keeping a northeastern course. When the depth had increased gradually from 6 to 25 fathoms, the *Colibri* signaled the *Berceau* which filled her sails. While still luffing the vessel the sea became very heavy, the weather did not look good, the wind was blowing hard and it was raining cats and dogs.

The last signals acknowledged by the crew of the *Colibri* were the order to reassemble in Mayotte in case of separation. The last time they saw the *Berceau* was when the three ships came about hard a weather close to Nosy Kalakajoro (the northernmost of the Radama islands) that they were trying to round. The bad weather worsened even more and the sea was furious. The *Colibri* came about several times during the night. At four in the morning the next day the crew realized that the *Voltigeur* had rounded the island. They were hoping to round it before long. A log and line was thrown. The *Colibri* was proceeding at two and a half knots but the dark weather made it impossible to see the island. A lookout was sent up to vigilantly watch for land. The brigantine sail clapped forcefully and threatened to tear up so two people were sent up to furl it. The *Colibri* was sailing on the starboard tack using the square topsail with two reefs, the courses and the inner jib. The ship was drifting heavily and the brutal pitch was putting a heavy strain on her.

The sky was pitch dark. When a few drops fell, sign of a possible squall four men were placed to haul up the lee-clue-garnet of the main sail and one man at the sheet. The watch was composed of only 15 men but two were off-service (the cook and one ill), two were manning the helm, two at the brigantine, one at the lookout and a Malagasy posted with the lantern leaving seven men available.

The safety measures concerning the main sail were taken a while earlier when a few raindrops fell and the wind freshened. The order was given to haul up the lee-clue-garnet. While this task was done however so slowly, the squall erupted suddenly and with immense force, causing the ship to pitch dangerously. This despite the luffing and striking the main topsail and ordering the same for the fore-topsail. In no time water passed over the rails and rushed into the ports. Though still luffing, the vessel stopped responding. Everybody was called on deck to get rid of the main topsail and the main-sail. But it was too late. The order was yelled to let go the topsail sheets. Unfortunately, this order was not followed and water started reaching the scuttles.

The helmsman called the Captain, who was below deck, but got no answer. With the ship keeled over the crew tried to scramble on her side. They again shouted for the Captain, but to no avail. The master of the yawl was busy cutting her tackle-falls when he suddenly felt the yawl sinking. With only part of the crew having jostled on her side the *Colibri* was swallowed up and disappeared.

The few survivors including Mr. Anquez started swimming. Someone grabbed Mr. Anquez' feet and both went under. He struggled to resurface but to no avail. The poor man, a Malagasy, gripped him with all his might and pulled him down. Three times, Mr. Anquez managed to resurface and each time he was dragged back down. Finally, after quite a battle, he freed himself and resurfaced. He found an empty cask, grabbed it and took some rest. Daylight was breaking and he came across another officer, Mr. Maureau, who was holding on to half the topgallant yard that had broken loose of the spar.

A little later they were joined by a third man, Lavaquaire, who was holding on to the stocks of one of the boats. After daybreak they were able to save another sailor, Cuviller, who was drowning. They picked up one of the masts of the long boat. After fastening together their spar with the shreds of Mr. Maureau's shirt all four carried on to the mainland, roughly four leagues away to where the sea, the winds and currents were pushing them. They later sighted seven other crew members holding on to an empty cask. As they were not moving at all, whereas our party was swimming towards shore they quickly lost sight of them. The squalls were less frequent but very strong. The torrential rains were extrememly cold and made them shiver. With their strength waning rapidly and being very tired they sighted a beach ahead of them where they seemed to be carried. At four in the evening, after taking some rest they ended up close to the reef and decided to split each taking part of the spar.

Figure 80. Sinking of the *Colibri* (Etch by Jules Noël in Zurcher, F. et Margollé, E., op. cit., p.173)

Mr. Anquez took the mast of the long boat and headed for land. Submerged by each wave that came crashing on his head he lost his spar, his only lifeline. He swam furiously to recover it. He found it and he ended up dry on the coral which tore his torso and arms. Panting and nearly powerless, he tried to stand up several times but the piercing pain caused by the coral gashing his feet thwarted that. He waited for the next wave to drag him along the bottom, causing excruciating pain. Finally, after two hours of inexpressible suffering he managed to land on the beach. Bleeding profusely and half-dead he took a few steps and fell down unable to go any further.

Suddenly he felt the water gaining ground. The tide was rising and wanted to seize its prey once again. His sheer weakness gave him little hope and night was falling. He tried an ultimate effort. Crawling on all fours he managed to lift himself high enough so as to be out of reach of the water. He remained there for a long while resting. Only when he saw the sailor Lavaquaire coming out of the bush close to him did he manage to stand up. He grabbed a stick and they walked together to a small watering place fifty feet further up. After bathing they spent the night under a tree shivering as the rain was still falling hard.

The next morning, they regained some strength and headed for Anorotsangana about one league from where they landed. On the way they met Cuviller who was with them on the spar and who was lucky enough to have made it. After they had split, he was the last one to leave Mr. Maureau who had told him that he was too tired and not to wait for him any longer; these were his last words. A Hova soldier seeing them suffer so much was kind enough to bring them to Mr. Renous, a Portuguese who lived among the Hovas a long time. He was most kind to them. There were already four other sailors at this house who had arrived the night before. These sailors, had jumped off the ship after it sank, found a piece of board and headed for the mainland. They landed around four in the afternoon without major injuries but being very feeble. They headed for Anorotsangana where they arrived around six in the evening and were hosted by Mr. Renous.

In the morning of February 28 they were told that the body of a white man was found on the beach. Taking linen and tools to dig a grave all seven went out to see who it was and administer last rites. They recognized Mr. Maureau who was not disfigured and did not seem to have drowned since he was distant from the high water point. He was covered in blood, pouring out of his nose, mouth and ears. He had neither cuts nor bruises on his body. Most likely, Mr. Maureau was hit by a violent wave on the stretch between the reef and the beach where they had all suffered so much. This wave probably thrust him to the bottom with such force that he fainted. Only being able to reach the beach during the night he died of want of help. This assumption was corroborated by the report of two Hova soldiers on guard duty about a hundred feet from where the body was found who claimed to have heard screams during the night. Mr. Maureau was interred and a cross made by one of them and bearing his name became his tombstone.

LE BERCEAU[186] (1846)

The *Berceau* was a small French navy vessel based in Reunion and used for protection of French interests in the southwestern Indian Ocean. She participated in various disturbances marking the reign of

Queen Ranavalona I. Since the 1820s, some European traders had established factories on Madagascar's east coast, particularly in the Tamatave area. They went quietly about their trade, avoiding at all terms to delve into the troubled politics of the time.

Out of the blue, Queen Ranavalona I issued a decree in 1845 summoning all foreigners living in Tamatave either to abandon their nationalities and become Malagasy, or to be gone within two weeks. The *Berceau* and the *Zélée* were in Tamatave's harbor, as was the British corvette *Conway*. The foreigners started embarking on them, while their properties were being plundered. In reprisal, the navy vessels bombarded the city and set it ablaze and some troops went ashore. Skirmishes ensued. Just before the vessels sailed off, the foreigners on board, on top of having lost everything, had to endure the gruesome sight of impaled heads of Europeans displayed on poles erected on the beach.

On December 13, 1846, a huge cyclone hit the *Berceau,* under the command of Captain Jean-Pierre Gout, while cruising off St. Marie Island. In no time, the ferocious winds and humongous waves engulfed her, carrying her down to the bottom of the ocean with all hands lost. The *Belle Poule* and the *Archimède* were sent from Reunion on a search party. The only traces they could find were some floating debris on St. Marie's northern and eastern coasts.

LAPÉROUSE[187] (1898)

The *Lapérouse* was a 200-ton French wooden hulled first-class mixed cruiser with iron beams laid down in Brest in 1875 and launched in 1877. Her crew consisted of 15 officers and 270 men and her armament of 15 cannons of 140 millimeters and some revolver-cannons. She participated in naval campaigns in China and the Indian Ocean then was assigned to the French colonial administration of Madagascar following its 1895 annexation. In 1897 General Gallieni, the French Governor, set out on a circumnavigation of the island on her in order to inspect the coastal settlements leaving Tamatave on May 19 and returning on June 27.

The next year, General Gallieni decided to undertake another circumnavigation of Madagascar. Having traveled overland from Antananarivo to Mahajanga where he stayed from June 16 through July 5, 1898 he boarded the *Lapérouse* under the command of Captain Huguet and escorted by the transport ship *Le Pouvoyeur.* He visited Nosy Be, Maintirano, Morondava and Tulear and arrived in Fort Dauphin on July 29.

The *Lapérouse* was moored in a depth of 33 feet using only her larboard anchor. Weather was nice on July 31, 1898 with a gentle northeastern breeze. The sea was rather rough with a heavy swell. At nightfall the winds suddenly freshened, sweeping into the bay with furious gusts, bursting the clouds and causing a massive downpour over Fort Dauphin. This first gust hove the sea which, whipped by successive squalls became heavier and heavier. The *Lapérouse* began to wallow without pitching too much. Unbeknownst to all, the cable of her anchor started to scrape the residue of the submerged wreck of the Norwegian vessel *Lotsen*. It was loaded with rubber and had foundered on that very spot four years earlier.

At 6:45 pm, at about time to order the clearing of the decks a sudden thump was felt at the bow. The anchor chain had ruptured. Pushed by the current and the strong wind the ship started drifting towards the coast with all her lights out because of want of coal. The officer on duty ordered the starboard anchor to be dropped. This operation took a few minutes. Before the second anchor bit the vessel moved back about 500 feet. Another thump was heard similar to the first. By astounding mischance the chain of the second anchor had also ruptured. The *Lapérouse* was now an inert mass at the mercy of the current and the gusts and swiftly ran aground.

During the night, Captain Huguet the officers and the men made several attempts to set up toings and froings with the shore. But the fury of the sea prevented them from doing so. The General, his officers and the population of Fort Dauphin assembled on the beach. They agonized the whole night fearing that the force of the storm would cause the vessel to capsize. At daybreak, the sea calmed a bit and the winds slackened. The damage could be assessed. The *Lapérouse* was beached in the middle of the breakers. The bow was sunk about six feet deep into the sand less than fifty meters from shore. Given her diagonal position the stern was somewhat less stuck. The absence of any other ship prevented any possibility to heave the cruiser off.

Organization for an orderly evacuation was set in motion. Without too much trouble, the crew was able to set up a to-and-fro motion with the shore through a slung cradle suspended to a ring gliding on a metal cable stretched between the bowsprit and the beach. The evacuation started immediately. The sick were offloaded first then the men. As much material as possible was brought ashore in particular; enough food to last a few days and all the firearms. Around three-thirty in the afternoon, the sea became heavier again and the Captain took the decision as a safety measure to clear the board before nightfall. A little before five all the men were safely ashore. Then, having ascertained that all was clear, the officers and the Captain last disembarked.

The Governor had to wait for the arrival of the coastal vessel *Tafna* which serviced secondary ports for the Compagnie Havraise Péninsulaire. He left on August 10 for Farafangana, then visited Manajary and Mahanoro before returning to Tamatave on September 3rd. The loss of the aviso *Lapérouse* was reviewed by the maritime war council in Toulon on November 4, 1898. Captain Huguet was acquitted as the council recognized that the events were beyond his control.

Figure 81. Postcard showing the wreck of the *Lapérouse* in Fort Dauphin

REFERENCES

[180] Sources: Wikipedia, "HMS Serapis" at http://en.wikipedia.org/wiki/HMS_Serapis_(1779); The Serapis project, at http://www.serapisproject.org/; s.n., "I have not yet begun to fight!" at http://www.hoala.org/Grade_7/jones.html; Correspondences kept at the Archives Nationales, Paris.

[181] The frigate *Bonhomme Richard* was originally called the *Duc de Duras*. Built in 1765 as a merchant ship for the French East India Company for service between France and the Orient, she was placed at the disposal of John Paul Jones on February 4, 1779 by King Louis XVI of France as a result of a loan to the United States by the French shipping magnate, Jacques-Donatien Le Ray.

[182] French naval archives (Services historiques de la Marine), Vincennes, Microfilm 0004, p. 46.

[183] Sources: Pfeiffer, I., "Voyage à Madagascar ," Paris: Hachette, 1881, pp. 25-26.

[184] Noted by Gerbeau, H., "Des minorités mal connues : esclaves indiens et malais des Mascareignes au XIXe siècle," in Table ronde sur "Migrations, minorités et échanges en Océan Indien, XIXe-XXe siècle," Sénanque, 1978, *Études et Documents,* Aix-en-Provence, IHPOM (Institut d'Histoire des Pays d'Outre-Mer), Université de Provence, n° 11, 1979, p. 160-242, at http://classiques.uqac.ca/contemporains/gerbeau_hubert/minorites_mal_connues/minorites_mal_connues.html (p. 20).

[185] Source: Report of Mr. Anquez to the commander of the naval station in Bourbon, April 23, 1843, copied in Zurcher, F., and Margollé, E., "Récits des Naufrages célèbres avant 1900," 5th ed., Paris: Hachette, 1888, Ch. 16, "Naufrage du brick le Colibri en vue des îles Radama (Madagascar) (1843)," pp.169-179; reedited Paris: La France pittoresque, 2000, Ch. XV.

[186] Sources: Pfeiffer, I., op. cit., p. 26-28.

[187] Source: s.n., "Voyage du Général Gallieni (Cinq mois autour de Madagascar)," Le Tour du Monde, 1899-1900, pp. 131-134; Bulletin du Comité de Madagascar, Paris, 1897, pp. 63-74 and 1898, pp. 456-460, 511-513, and 635-636 ; Musée Lapérouse, Albi, « Découvrir, Navires en service dont le nom évoque Lapérouse », at http://www.mairie-albi.fr/arthisto/gens/navires.html.

8

Sailing Vessels and Steamers of the 19th Century

The 19th century saw the transition from the graceful sailing craft to passenger and package freight steamers. Though sailing ships underwent significant changes including: hulls built of iron (then steel), masts of steel, and using wire and chain instead of hemp rope, the efficiency and growing reliability of steam-powered vessels hastened the end of the age of sail. Madagascar also saw remarkable changes during that century. The Merina kingdom unified the country and it was followed by French annexation late in the century. Tamatave became by far the country's main harbor. As a result, nearly all wreckages of the period happened on Madagascar's east coast.

ELISA[188] (1827)

The *Elisa,* a British owned bark operating in Mauritius, had taken several voyages to Madagascar since 1820. In the early morning of November 26, 1827 under the command of Captain Lenepveu and full of merchandise, she left Tamatave's dock bound for Mauritius. A southwestern breeze facilitated the maneuver. Just after exiting the

Figure 82. The Chilean training vessel *Esmeralda* ("the White Lady") docked in Tamatave (reproduced with the kind permission of Mr. Jean Philippe)

pass and rounding the reef on the starboard side towards the open sea, the wind shifted to the southeast and a strong northern current pushed the ship towards the northern reef. Orders were given to tack. When starting to veer a slight jolt was felt as the *Elisa* struck the reef. Soon after, the helm was ripped off when it hit a rock slightly seaward of the reef. The pounding surf caused the unsteerable ship to crash on the reef where she broke up. Ten minutes later water had reached the steerage level. The crew cut down the masts to keep her up and off-loaded as much cargo as possible. Barges were sent from the harbor to pick up the passengers and the cargo. The *Elisa* then broke apart with a dreadful crash.

MARGARET OAKLEY[189] (1837)

The voyage of the brig *Margaret Oakley* is among the late journeys of the American sealing Captain and explorer Benjamin Morrell. He was known mostly for his early sealing expeditions in the South Seas.[190] He brought two savages from earlier expeditions; Sunday and Monday, back to the United States and displayed them in an exhibition in New York. He wanted to teach them English and use them as interpreters for future communication with their native lands when he returned. Unfortunately, Monday died from homesickness and Captain Morrell suspended his scheme. Soon after, three gentlemen offered to sponsor an expedition for Captain Morrell. But on a less expensive scale than the one he had planned.

A new hull of a symmetrical clipper brig of 230 tons, lying on the stocks on the eastern shore of Chesapeake Bay, was to be purchased, launched, rigged and sent to New York to receive her trading cargo and general outfit for the voyage. On a pleasant March evening at about sunset, Captain Morrell stepped into a small boat on Castle Garden. The boatman plied his oars and he waved a long farewell to his native city. The rising of the sun on Sunday, March 9, 1834 saw him on board of the brig *Margaret Oakley* lying at anchor in the Hudson river. The wind blew strong and keen from the northwest. Anchor up and away!

At 4 pm, the *Margaret Oakley* was fairly out at sea on a three years' South Pacific exploration cruise. After crossing the Atlantic Ocean, the ship stopped at the Cape Verde Islands for provisions then headed south. She passed the Tristan da Cunha Islands, a group of remote volcanic islands in the south Atlantic Ocean, and attempted to head for the South Pole. However, the *Margaret Oakley* was caught in a storm and ended up in Mauritius in June 1834. She underwent repairs during the next three months. On September 2, she sailed due east into the Indian Ocean. By October 3rd she dropped anchor at Sandalwood Island (now called Sumba in Indonesia). Captain Morrell briefly explored the Sea of Celebes, the Banda Sea, Cajeli Bay, the Islands of Bomboa and Manipa, the Gause Straight, Gillolo Island, Geby and Jeoy Islands, Waigoo Island and the Youle Islands—at the time unknown and unexplored regions of Tropical Australasia—before reaching Papua.

The *Margaret Oakley* then headed northeast from Papua into the Pacific Ocean. Captain Morrell spent quite a lot of time discovering and exploring a group of islands named after him (Morrell Islands, now also called Thule, the southernmost islands in the South Sandwich Islands). He then cruised on a southward course through the South Seas, undertaking original exploration of many islands and archipelagoes. Not appreciating the names given to the islands of tropical Australasia by the British, which he felt were "inappropriate and ignorantly applied," he decided to rename them after their native names which were more fitting to the climate and the geography of the place. So, New Britain became Bidera; New Ireland became

Figure 83. The *Margaret Oakley* in Papua (Source: Jacobs, T.J., op. cit., p. 73)

Emeno; New Hannover, Pelego, and the Central Island and Admiralty, Marso.

In Bidera, Captain Morrell and the crew, after a few days of trying to gain the natives' confidence, were welcomed by them and had a great relationship with many of the village chiefs. Thomas J. Jacobs and his friend Selim E. Woodworth, a young man who had had an adventurous youth and spent most of his time at sea, learned the language and became known as "medicine men" in the island when they managed to help a few of the locals with western medicine they had aboard. Cleverly, they had acted out a tribal ritual to give this medicine to the villagers. When it was time to leave the Bidera Kings gave Captain Morrell a present of ancient skulls as a sign of sacred bond.

The *Margaret Oakley* sailed to Australia and dropped anchor in Sydney in June 1835 at the beginning of the winter. The crew was used to the heat of the tropics by now. They became cold and ill in the Australian winter. Captain Morrell decided to quickly head back to Papua to explore the islands around there. He sailed west to Borneo, explored the Sooloo and Mindoro Seas and a number of islands and anchored for a while at Manila Bay (Philippines). Then he headed to the China Sea and Canton Bay where the *Margaret Oakley* anchored at Wampoa. There, Captain Morrell announced that the romance of the voyage had ended and the ship was now merely a merchant vessel bound for New York with cargo.

Shortly after leaving for home the *Margaret Oakley* was caught in a typhoon in the China Sea which caused her to loose the rudder. She wandered with a temporary rudder for a while until calling at Singapore to undergo repairs. In late 1836, intelligence was received in New York of her departure from Singapore but no one knew her subsequent course. But she still had not arrived. There were many idle rumors afloat among the merchants and insurance officers. One asserted that Captain Morrell had turned pirate, another that he had run away with the cargo and had formed a settlement upon some island in the Pacific Ocean. There was a whisper that she might have foundered.

In fact, after a few weeks undertaking repairs in Singapore, the *Margaret Oakley* set sail with only 35 days provisions on board. She passed through the Straits of Banca (Selat Bangka) and anchored one day at Mintow. After this, she ran on a mud bank and got off after three hours' hard toil. Passing the Straits of Sunda, she shaped her course westwards over the Indian Ocean for the Coco's Islands. Short of provisions, the crew was put on a per diem allowance of half a pound of meat and half a gallon of water. She passed the Coco's and six weeks after leaving Singapore was anchored at Fort Dauphin on Madagascar's southeast coast.

On the morning of the third day after the *Margaret Oakley*'s arrival, while the Captain and part of the crew were on shore buying cattle, a gale wind arose and a dangerous sea set on shore. The vessel plunged and staggered. The waves completely buried the bow, sweeping the deck. The anchor dragged and another was let go to no purpose. The cables parted and she drove broadside before the gale. Captain Morrell arrived upon the beach just in time to see the vessel drive on shore and bilge. In half an hour there was three feet of water in the hold and it was increasing incessantly.

By this time, 200 natives had collected upon the beach. All were willing to lend a hand in "saving" the cargo. A string of natives was soon formed reaching from the vessel to the shore some in boats and canoes and others standing in the water. It soon turned into chaos with the natives opening the boxes and running away with all they could carry: silks, satins, crepes, costly shawls, boxes of tea, curious productions of China, all going. Captain Morrell went insane. He held out his gun and threatened to shoot whoever opened a box. But as soon as he turned his back, open went a box and away ran the natives with the contents.

The crew managed to reach shore safely and was put to guard the cargo. The next day it was secured from the weather with tents made of sails. The vessel became a total loss. Different articles drifted ashore from the wreck. That day the natives found the skulls of the Bidera Kings that Morrell was carrying aboard in one of the boxes. From that moment on, the natives thought the crew of the *Margaret Oakley* were a set of piratical cannibals who were cruising the shores of Madagascar eating the people and preserving their skulls. The natives were furious and ready to attack. Captain Morrell had to use great tact to calm them down.

The crew stayed in Fort Dauphin for several months. Two black members of the crew became great favorites with the natives of Fort Dauphin. Each built a house, a garden and married Malagasy women. They ran a business salting beef and curing hides. Captain Morrell eventually managed to send news of the loss of the *Margaret Oakley* to Mauritius. The American consul there immediately dispatched a vessel to the relief of the crew. Some of them were taken on board with part of the cargo. Thence, Captain Morrell and the remainder of the crew and cargo were taken by a British vessel and arrived safely in England. Captain Morrell then sailed to the West Indies where he prepared a clipper vessel to sail to the east again. In 1839 he died of the "prevailing fevers" in Mozambique.

While the crew was lingering in Fort Dauphin, Selim E. Woodworth started an inland journey through Madagascar with the intention of reaching Tamatave. He was captured by natives who threatened his life and took him inland where they kept him as a prisoner on an islet in the centre of a large lake. At night crocodiles and other creatures of the water crept over the islet and forced him to sleep up in a tree. He was kept there for two days without food and fearing for his life every moment. On the third day, a Malagasy girl on a canoe who took pity on him brought him food. Woodworth then moved to the village and lived for a while among the natives. A few months later the native woman helped him escape. He got away on a whaler and reached home after having been given up for dead. He entered the Navy as midshipman in 1838.

INDIENNE[191] (1852)

The corvette *Indienne,* a French transport ship commanded by Captain Proté, set out from Bourbon in the middle of December 1852 bound for Mayotte in the Comoros islands. The wind was fair and the sea calm, promising a safe voyage. At the extremity of the narrow channel separating St. Marie's Island from Madagascar's main land, on doubling the bay of Antongil, the sea became more boisterous and the vessel began to pitch. The night was very dark, the wind blew a perfect gale and the sea rose in proportion. The sails were reefed one after another. The only two that remained were carried away by a sudden squall. The ship leaned on her side as if she had struck and water rushed into the hold. At the same moment, with a dreadful crash, the main and fore-masts both fell by the board, crushing two of the boats. The main mast fell inboard and lay across the deck.

The falling masts enabled the vessel to right herself. Water rushed in at intervals. The *Indienne* sailed onward during the night carried along by the current, but without any power to direct the ship's course. The commander ordered all hands on deck. It was after eight o'clock in the morning but the sky was obscured. The rain fell in such torrents that one could not see beyond the vessel and the deck was washed at every instant by the waves. After a meeting, the officers came to the unanimous decision to run the vessel ashore rather than founder in the open sea. A sudden glimpse of light enabled them to sight the land which was not more than two miles distant.

But a long line of rocks lay concealed very close by. As shipwreck appeared imminent and as a last attempt to save the ship, Captain Proté gave orders to put her about. But she no longer answered the helm. The *Indienne* consequently drifted at the mercy of the waves. Some of the seamen, with a view to lighten the vessel, dismounted the pieces of artillery and threw them into the sea while others opened the port-lids to obtain a free outlet for the water. These precautions were necessary as the ship was within a short space of the rocks. Two strong waves soon drove her upon them.

At this most critical and perilous moment, the Captain exclaimed "All hands astern!" There was an immediate rush in that direction. Another breaker, stronger than any experienced before, struck the stern of the *Indienne* which suddenly changed position. She then drifted over the reef at a point where it was somewhat lower, but which would have been insurmountable without the tremendous wave lifting her above it and throwing her into the calm sea beyond. The corvette continued her course avoiding every shoal and discovered an unknown channel. She sailed down this. On arriving at the end she rested calmly on a solid bottom where she could set at defiance to the remainder of the storm and resist its fury.

The gale continued to rise. The waves acted with such violence on the masts and rigging that the ship was exposed to considerable danger. With a view to save her, the rigging was cut away and thrown into the sea. Towards evening the weather became calmer and the crew and passengers could see the land. They were at a about 70 miles distant of St. Marie, a little to the northward of Antongil Bay and close to Tanjona. The whole coast bore evident marks of the devastation committed by the storm. Trees were torn up by the roots and houses blown down in every direction.

As soon as it was possible one of the two remaining boats was put to sea. Well armed and commanded by an officer it steered to the shore where the men were well received by the Hovas. However, they refused to give any provisions or even fresh water. These were the rigid orders of queen Ranavalona with regard to all strangers even after a shipwreck. Fortunately, the *Indienne* had plenty of stores and two cannons and other arms secured her against an attack. Next came the question of how to get the corvette afloat. This required the assistance of another vessel. Perhaps the *Caïman* or the *Victor* were possibly still anchored in St. Marie Island. It was resolved that the longboat should be sent off as soon as possible.

Early the next morning, seventeen men left with the long-boat, taking a compass and provisions for five days. As the wind was very low the men had to row the boat most of the day. The first night was dark with frequent squalls. With the compass more necessary than ever they were alarmed when they perceived that, due to the rain that fell on it, it no longer pointed to the north. To crown their misfortune, the lamp doused off and the current drifted them off their course. The next morning, their position became more alarming in proportion as they advanced and nothing indicated any improvement in the weather. Fortunately, later in the day a fresh breeze suddenly arose from the most favorable point. They were able to unfurl all their sails. Directing their course toward St. Marie, which they perceived in the distance, they arrived at the coast in a short time.

They soon saw the *Caïman,* sailing directly towards them. Seeing their signals the crew of the *Caïman* stopped the engines surmising that the *Indienne* must have been wrecked and that this small party alone escaped. On arriving on board, the men explained their misfortunes and unexpected deliverance. Orders were given to put about and return to the port of St. Marie. The next morning the *Caïman* got under way and sailed towards the *Victor,* which she took in tow. The two ships proceeded together to the *Indienne* which they reached towards the evening of the following day. She appeared in a deplorable condition but the greatest joy was manifested by the crew on meeting together.

The salvage work commenced at once. Three days were spent in sounding and marking out a free passage and in discharging the corvette. Anchors were cast in different directions so as to approach the stranded vessel as nearly as possible. On the fourth day the powerful engines of the *Caïman* were set in motion. In spite of the force exercised upon the capstans and by the 250-horse power engines, the *Indienne* was at first immoveable. But in a short time she began to move, was gradually brought into line and at length towed into the channel. The Hovas, astonished at this hazardous feat, came on board the *Caïman* to congratulate the Captain on his success. The rest of the day was spent in fishing up here and there and collecting the pieces of masts and rigging that were lost. In the evening, the three vessels set out on their return to St. Marie where they arrived about the middle of the following day.

MACASSAR[192] (1880)

The 259-ton French brig *Macassar*, built in 1872 in St. Malo, sailed under the command of Captain Casimir Wilhelm since she was launched. For her last fateful voyage she was fitted out at St. Denis (Reunion) on June 16, 1880 for a commercial run to Madagascar. She had a crew of nine, both French and Malagasy, and Mr. Godré, a Mauritian master, embarked on her as a passenger. Loaded with a cargo of rum and salted beef the *Macassar* weighed anchor from Tamatave on August 7, 1880 at 8 o'clock in the morning. She exited the harbor's large channel with fairly good weather and light winds and shaped her course eastnortheast bound for Antongil Bay to fetch a load of rice that had been arranged beforehand. The scheduled route called for rounding St. Marie Island to the east and then to sway on a northwestern course towards the entrance to Antongil Bay.

At nine thirty in the morning, the *Macassar*, cruising at around five knots, was about three miles at sea opposite Prunes Island. Then things started to deteriorate. The skies darkened, squalls inundated the deck, and the sea became heavy. The Captain steadfastly kept his course to the eastnortheast on his compass, confident that this would steer him well clear of the reefs of St. Marie Island. He nevertheless ignored the very strong currents often running between Tamatave and Antongil and a concomitant westward drift. Also, he did not bother regularly throwing the log and line to check his speed. In the afternoon, fog settled in, the skies darkened even more, and the rain intensified. The first officer, Léopold Ramel, took the helm at six in the evening. Winds were blowing from the eastsoutheast. He sailed nearer to the wind at about four knots, and placed a man at the davit with orders to carefully look for reefs or passing ships.

The Captain took back the wheel at eight in the evening. Still confident to be well out at sea, he did not deem it necessary to send up a lookout nor to sound. Even so, Antoine, a Malagasy sailor, was apprehensive. On his own volition, he climbed at around nine in the evening in the masts, but could not discern anything. At the same time, Pierre Bébat, a novice posted astern, faintly caught sight of a white line; it was not very clear-cut, though, and he refrained from sounding the alarm. At 9:20 pm, the Captain passed on the helm to Antoine and went down to his cabin. A few minutes later, Antoine, more and more worried, suddenly asked the novice Pierre Bébat to take over the wheel and he rushed back up the masts. Now clearly seeing white waves, he yelled "Luff, there is a reef larboard!."

Immediately the Captain climbed back on deck and also clearly saw the white line on the port side, very close, and which seemed to pull out in front of the ship. He ordered the other watch to keep the luff nearest to the wind, but with slack winds coming from the eastsoutheast and heavy seas, he realized that there was no way that he could round the reef. Knowing that the vessel would not put about head to wind, he decided to heave astern and ordered to brace about the yards. With the winds that slack, the vessel moved ever so slowly, even with all the sails braced nearest to the wind on the port side. She was luffing, though, and crew thought they were nearly saved, when suddenly two or three huge groundswells took her broadside and heaved the *Macassar* against the reefs of Blévec Point, the southernmost point of the Ile aux Nattes. She immediately touched ground. The floodtide and heavy swells pushed her more and more inland. The fury of the surf was such that nothing could be done beyond preventing the crew from getting hurt. The Captain ordered everybody on the poop, where they spent the night being drenched by the sea. At eleven, the masts fell off and the vessel broke up.

The next morning, the water had retreated with the ebb tide and the *Macassar* lay nearly dry. They noticed that the brig was lying on her starboard side, that part of the bridge was demolished, that all of the masts were knocked down, and that she was rammed in several places. She was lost, and the violence of the surf in that spot would soon completely break her up. The chief and inhabitants of a nearby village soon turned up in dugouts. The crew managed to swing out the yawl. Having offloaded the ship's money chests and a few instruments, the crew and the passenger sailed to Madame islet of St. Marie Island, leaving behind a sentry designated by the village chief to look after the cargo.

The ensuing investigation concluded that Captain Wilhelm's steering had been negligent and that the crew lacked the required skills and experience. In particular, the Captain should not have relied solely on his compass. He should have assumed that the often strong currents in these waters could veer off the vessel's course westwards, and hence should have followed a more prudent route.

OISE, SARAH-HOBART, ARGO, CLÉMENCE AND ARMIDE[193] (1885)

In June 1863, Queen Rasoherina succeeded her husband, Radama II who was assassinated by military officers. Her new husband, Prime Minister Rainilaiarivony, exercised the real power in the land. In 1868 Queen Rasoherina passed away. He elected Queen Ranavalona II in her stead and married her without more ado. Under his iron grip, the kingdom got organized and developed rapidly. This autocratic Hova, driven by nationalistic feelings, gradually started to rescind the old treaties, in particular the right of foreigners to own land which had been granted in 1865. In 1880 he seized the land and the house in Antananarivo of the former French consul Laborde who passed away two years earlier. This was too much for the French to stomach. In 1883 an expedition was sent under the command of Admiral Pierre. They bombarded and then occupied the two main harbors; Mahajanga and Tamatave, for the better part of two years. In the meanwhile, Queen Ranavalona II died in late 1883 and was succeeded by the young Queen Ranavalona III. In 1884, Admiral Miot replaced Admiral Pierre to lead this occupation and the occasional inland skirmishes with Malagasy troops.

In the middle of this French campaign Tamatave was hit by a major cyclone on February 24, 1885. The French transport ship *Oise* had just arrived from Europe. She was lost south of hotel Lagrave. Twelve crew members perished in the accident. Just in front of hotel Lagrave, the French liner *Argo,* operating the annex cruise line Reunion-Madagascar-Comoros, which was full of provisions for the French troops, was cut in two. It took significant trouble to save her crew. The same storm also claimed the American bark *Sarah-Hobart*[194] commanded by Captain Crocker, at Tanio point; the French schooner *Clémence* thrown ashore just north of the *Argo;* and the Mauritian vessel *Armide.*

SURPRISE[195] (1885)

The 473-ton American bark *Surprise* was built in 1866 at East Boston. Her principal owner was B.F. Hoyt of Boston. On August 16, 1885 she left New York harbor under the command of Captain Cyrus B. Averill loaded with 16,500 cases of refined petroleum bound for Bombay via Zanzibar and Chittagong (Bangladesh's chief seaport). After an uneventful crossing of the Atlantic, the *Surprise* took the inner passage. She was working up Madagascar's western coast endeavoring to keep in shore of the strong southerly current and obtain favorable land winds. At 1:30 am on November 21, 1885 with clear weather and bright moonlight, Captain Averill judged to be about 30 miles off-shore. He tacked in-shore and went below ordering to be woken up if land was in sight.

At day break, soon after 4 o'clock, the vessel suddenly took bottom. The man at the wheel immediately hove the wheel down but the second officer in charge of the deck, ran aft and inexplicably hove the wheel hard up—the vessel still going ahead. Captain Averill ran on deck and ordered the wheel to be hove hard down. But the vessel's headway being nearly done it had little effect. She stopped almost immediately thereafter. The *Surprise* was stranded on a reef lying about three miles off the coast near Salara on the southwest coast.

The wind blew fresh from the northeast all day nearly off-shore. As Captain Averill found it to be about one hour ebb tide when she ran ashore he thought that he should be able to get the vessel off on the next high water. They clewed up sails, got a boat out and ran out the stream anchor and hawser after which they jettisoned cargo. They made every effort to get the vessel off but with no success. The *Surprise* rested nearly dry at low water and thumped heavily in the reef on the flood tide. She began to leak. Captain Averill gave up all hope of getting the vessel off as he could get no assistance from shore. The natives were alongside in canoes during the day appearing friendly. But they offered no assistance. After the ebb tide, the crew manned the pumps and worked until they were beat-out. Even then they found they had three feet of water in the hold.

They determined to abandon ship in the morning while the weather was favorable deeming it useless and unsafe to remain longer on board. At 3 am on November 22 they began to make preparations to abandon ship—the rudder was broken by then and the water in the hold had reached seven feet. They put the boats in the water, putting spars, sails, rigging, water and such stores and clothing as they were able to take in the long boat. They sent it inside the reef in charge of the first mate G.B. Crocket and two men. Then they loaded the other boat with their trunks, clothing, the chronometer, barometer, compass and nautical instruments. While doing so the long boat was surrounded by natives in canoes who showed no violence. Two of the natives without arms boarded the long boat and the canoes came alongside the vessel.

Captain Averill left the ship soon after with his daughter and the rest of the crew and pulled into the long boat. The natives remained alongside while some of them went aboard the ship. The two natives in the long boat wanted them to go ashore pointing to a heavy squall making up in the southwest. Captain Averill refused to go ashore, making them understand that he was going south having seen vessels at Nosy Ve when passing that island. After pulling in shore under the lee of a small point, they moored the boats and lay until the squall passed over. The sky having cleared, one of the canoes paddled close to them and several others stacked in from the vessel. The crew of the *Surprise,* wishing to cast off, tried to get the two men out of the long boat into the first canoe before the others came but could not do so without using violent means. Captain Averill did not deem this prudent under the circumstances.

The canoes came in rapidly. Suddenly the two natives left the long boat. As the crew attempted to proceed, they became entirely surrounded by 25 or 30 canoes with 3 to 4 natives in each armed with guns and spears. Seizing their spears, they jumped into the Captain's boat sinking her to the thwarts and taking everything out of her. They even grabbed an oil coat and shawl off the Captain's daughter's back. They left the provisions and water in the long boat and allowed the crew of the *Surprise* to proceed on their way south.

The following morning, November 23rd they sighted a bark at anchor in Tulear Bay. They ran alongside, made fact about noon and were kindly invited on board. The vessel was the French bark *Notre Dame de la Garde* commanded by Captain Bellard who treated them very kindly. They then went ashore and made contact with the agents of a Port Natal (Durban) coastal trading company. Captain Averill arranged assistance with them and made provisions for the crew which was sent to Nosy Ve in the long boat to await passage to Port Natal.

Leaving his daughter in Tulear, Captain Averill returned the next day, November 24 to the wreck of the *Surprise* accompanied by one of his men and the agents of the Port Natal trading company, arriving that evening. The following morning they tried to board the wreck, but could not do so on account of the sea which was breaking over the sail. Even if they had been able to board it would have been of no use as the natives would not have allowed them to take anything without force. In the meantime, the natives had taken out a large part of the cargo of oil, stripped off the metal, cut the sails and rigging in pieces to make division of spoils and had burnt the hull to the water edge.

Plundering the boats while afloat was undoubtedly an act of piracy. It was done by members of the Sakalava tribe. In those days the Hova government in Antananarivo, lead by the young Queen Ranavalona III and her husband Prime Minister Rainilaiarivony, were unable to cope with a state of anarchy and disorders on the west coast. The region swarmed with Sakalava tribesmen who paid only nominal obedience to the central government and a weird assortment of cutthroats and slavers who both preyed on and used the Sakalavas in their ruthless schemes.

Captain Averill made up a list of the things stolen from the boats and sent it to the local King Lahamarisa asking for his assistance in their recovery. The King replied that he could do nothing unless he knew who the actual parties were who took the things. The situation became more complicated when it was learned that the commander of the customs house at Ranopas, in company with a certain Hassan Ali, had purchased a large quantity of the oil that formed part of the *Surprise*'s cargo from the Sakalavas who plundered the ship. They had shipped it away. The Secretary of State of the US government was officially informed, protests were lodged and during the next three years, the American government and its consular officer in Andakabe (Morondava) Mr. Victor F. W. Stanwood, asked redress for the outrage and the victims to no avail.

In late 2007 Fred Lucas and this author discovered a wreck in Salara which could be the remains of the *Surprise*. The wooden hull and its copper lining are still in good state and the outer planks seem to have suffered from fire. There are three large anchors at the site. Several pulleys, copper nails, pieces of raffia hawser and other artifacts were found.

Figure 84. The main anchor

Figure 85. Planks from the hull

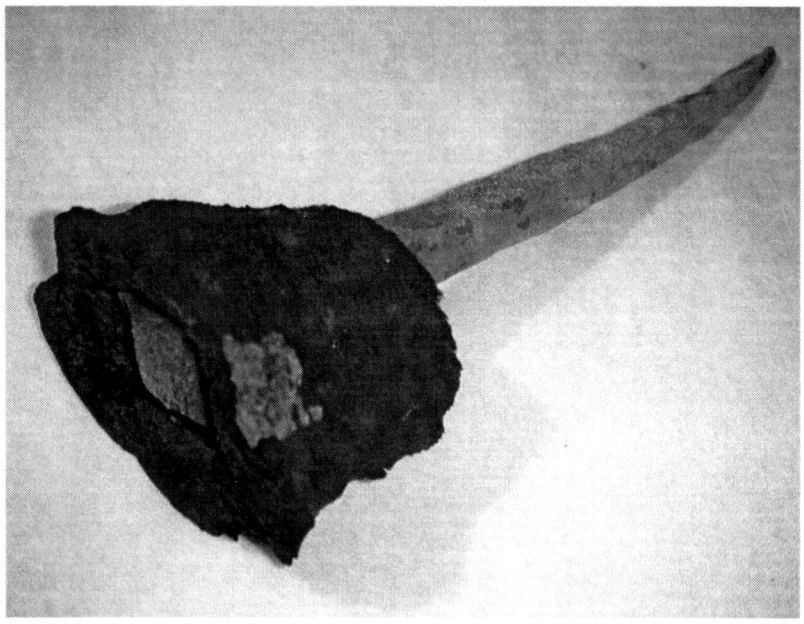

Figure 86. Large copper nail

Figure 87. Glass container

Dayot, Belette, and Glide[196] (1888)

On Wednesday, February 22, 1888 Tamatave was hit by a violent cyclone. Though lasting only from 11 in the morning to around five in the evening the damage was considerable, particularly in the harbor. The French state cruiser *Dayot,* notwithstanding the efforts of Captain Daniel and his crew to save her, was thrown on the large reef right off Tanio point and foundered. The ship was broken in two. The forward part submerged and the stern sat on the reef. One life was lost when the panicked sailor Coeffic jumped overboard to try to swim to shore but got carried at sea and drowned. The rest of the officers and crew were safely brought ashore. The sinking of the *Dayot* was no great loss in that she had been ordered to return to France to be put out of commission. Her boilers were so old that she reached a maximum speed of 5 knots. The English schooner *Belette* sank at the same spot. The American bark *Glide* foundered at the southern point of the large reef with no loss of lives. A number of small coastal ships and boats were also lost during the same typhoon.

Figure 88. The French ship *Dayot* and the British schooner *Belette* sunk in Tamatave harbor after the cyclone (Source: La Nature, 1888, p. 317)

SOLITAIRE[197] (1888)

The wreckage of the American schooner *Solitaire* is intractably linked with the tragic murder of Victor F. W. Stanwood, the American consular officer at Andakabe (Morondava). He had vociferously tried to obtain redress for the victims of the bark *Surprise* (see above).

Mr. Stanwood, a former sea Captain reportedly born in Boston, arrived on the southwest coast of Madagascar about 1877 and established a trading business at Morondava then called Andakabe. He was reported as a "windy American" who was putting on airs to amuse himself or gain prestige with the Sakalavas—a tribe inhabiting that part of Madagascar. The American consul at Tamatave, W. W. Robinson, was alerted in April 1880 when he heard that a man signing himself U.S. Vice-Consul had written letters to the capital Antananarivo from the west coast complaining of outrages, notably traffic in slaves and contraband (particularly arms and gunpowder) committed there by the natives and a band of lawless men of various nationalities.

As mentioned above, the Hova government in Antananarivo had little effective control over the dealings on the west coast at the time. Slave dealers operated with barren impunity in collusion with the local provincial authorities. They brought African slaves into Madagascar and exported Malagasy natives for sale up and down the coast of Madagascar and transported them to points of resale in Africa. Stanwood was particularly infuriated because the slave traders and other dealers in contraband were deceitfully flying the American flag to which they had no claim. It was an insurance policy against search and arrest. He also spoke of British subjects masterminding the traffic in slaves and contraband and accused the British consular officers and the whole British community on the island of scheming to undermine America's rather considerable commercial interests on Madagascar's west coast.

Nonetheless, the need for an American representative on that part of the island was deemed important. Notwithstanding reports branding Stanwood as an unreliable adventurer, he was labeled as "energetic and driving." So he was appointed by Consul Robinson on October 12, 1881 to be United States consular agent at Andakabe. He continued to keep a voluminous dossier on the brigands and slave traders and their powerful accomplices in high places. And he was a prolific writer of indignant correspondence often complaining that his communications went unanswered and that there was a total abandonment of all efforts to further American interests.

The bitter climax to Stanwood's efforts to see justice done to American interests on the turbulent west coast of Madagascar came in the fall of 1888. In September of that year an American-owned schooner, the *Solitaire,* foundered at the port of Belo-sur-Mer. Its Captain was one L. du Vergé, who had served in the Malagasy forces during the Franco-Malagasy (Hova) War of 1883-1885. du Vergé claimed to be an American citizen but Stanwood's investigations revealed him to be citizen of Mauritius. Stanwood also reported that the *Solitaire* plied the west coat for months and he strongly suspected that it had dealt in powder, munitions and possibly slaves. The newly-arrived American Consul, John. P. Campbell later reported that the *Solitaire* beyond doubt was carrying a cargo in part made up of arms and ammunitions—importation of which into Madagascar by American citizens was forbidden by the US-Malagasy treaty of 1881.

Du Vergé was a nineteenth century swashbuckler of sorts, a charming pirate, who moved with ease in respectable circles. Various reports label the relations between Stanwood and du Vergé as amicable—some state that they were business partners. However, when Stanwood ascertained the character of the cargo he changed his attitude toward du Vergé. After the wreckage Stanwood went to Belo-sur-Mer twice on behalf of complaints by the seven American seamen claiming wages due to them but could not meet du Vergé and had to leave without satisfaction.

On November 5, 1888 following another complaint by the crew of a fight with the Captain, Stanwood once again returned to Belo this time in the company of some Hova officers. He went to Captain du Vergé's rented house. That day there was already an episode of violence in the house. One of the American seamen, William Simmons, had come to blows with the Captain, in the process of demanding his back wages. He was driven off when du Vergé's brother-in-law began shooting his pistol into the air. So the atmosphere was highly charged when Stanwood arrived purportedly to make an arrest. Captain du Vergé ended their brief argument with revolver shots fired at close range into his chest, instantly killing the U.S. consular official.

SS ASIATIC (AMBRIZ)[198] (1903)

SS *Asiatic* (sometimes operated as the RMS *Asiatic*) was built as a cargo vessel in 1871 at Liverpool by Thomas Royden & Sons. She was typical of the transformation from sail to steam with the capability of both having three masts and one steam-driven propeller. She weighed 2,122 tons and had a service speed of 12 knots. Launched on December 1, 1871 she was built 'on spec' and purchased by the Oceanic Steam Navigation Company while she was being fitted out. In March of 1872 she was placed on the unsuccessful Calcutta trade. Later that year she went on the equally unsuccessful South America route although her first voyage was on charter to Lamport & Holt. On February 25, 1873 she commenced her first voyage to South America for the White Star Line but it was not profitable.

In 1873 following the wreck of the *Atlantic* off the coast of Nova Scotia, *Asiatic* and her sister ship *Tropic* were sold by the White Star Line to recoup capital. *Asiatic* was sold to South African Steam Ship Company later to become the Elder Dempster Lines. She was renamed *Ambriz*. Their largest ship at the time, she commenced her first sailing to West Africa on September 12, 1873. In December 1883 she was refitted, reboilered and placed on the Liverpool to New Orleans cotton run the following year. In 1896 she was sold to Cie. Française de Charbonnage et de la Batelage of Mahajanga. She was deployed as a mobile coal depot ship which steamed to Europe, usually Cardiff when stocks needed replenishing. She was wrecked in February 1903 off the coast of Madagascar. There were no fatalities.

PESHAWUR (ASHRUF)[199] (1905)

Peshawur, a 3,782-ton iron mixed passenger-cargo steamship was built by Caird & Company at Greenock in 1871 and powered by a 2-cylinder compound inverted steam engine. She was owned by the Peninsular & Oriental Steam Navigation Company (P&O Line) and was among the first P&O ships designed for the Suez Canal transit. She plied the Southampton/Bombay, Far East or Australia route carrying 164 First class and 53 Second class passengers. She was the 50th P&O Line ship to visit Australia, making return voyages in 1881 and 1882. In 1896, she was employed on Indian trooping. In 1900 the *Peshawur* was sold to Hajee Cassum Joosub from Bombay for pilgrim trade to Jeddah and renamed *Ashruf*. She foundered at the entrance to Tamatave harbor on May 8, 1905, when on a voyage from Port Louis, Mauritius to Tamatave.

Figure 89. The *Peshawur* (source P&O Lines Heritage Collection)

Figure 90. View of Tamatave around 1900 (reproduced with the kind permission of the Bibliothèque Grandidier, Antananarivo)

Figure 91. View of Tamatave around 1900 (reproduced with the kind permission of the Bibliothèque Grandidier, Antananarivo)

Figure 92. View of Tamatave around 1900 (reproduced with the kind permission of Mr. Jean Philippe)

REFERENCES

[188] Source: Valette, J., "Le naufrage de l'Elisa à Tamatave, le 26 novembre 1827," Bulletin de Madagascar, No. 254-255, July-August 1967, pp. 533-536.

[189] Jacobs, T. J., "Scenes, Incidents, and Adventures in the Pacific Ocean, or the Islands of the Australian Seas, during the cruise of the clipper *Margaret Oakley*, under Capt. Benjamin Morrell," New York: Harper and Brothers, 1844.

[190] Recounted in his "Narrative of Four Voyages to the South Seas," New York: J. & J. Harper, 1832.

[191] Source: Relation of Rev. Cotain, Jesuit Missionary at Madagascar, in "Tales of Shipwrecks and Peril," London: Burns and Lambert, 1858, pp. 76-85.

[192] Source: various documents and testimonies, Service Historique de la Défense, Paris, N° CC-4-2159

[193] Sources: Le Chartier, H. and Pellegrin, G., "Madagascar depuis sa découverte jusqu'à nos jours," Paris: Jouvet, 1888, pp. 118-119; Chauvin, J., "Le Vieux Tamatave (1700-1936)," Tamatave: F. Sourd, 1945, pp. 136-137.

[194] Also called *Sarah-Burk*. Built around 1860, originally from Freeport, Maine, and later registered in Boston, she had plied the Indian Ocean to New York route for more than twenty years.

[195] Sources: Various reports and correspondences, Archives Nationales de Madagascar, Antananarivo, Series DD.16 and DD.17; The New York Times, March 15, 1886.

[196] Sources: s.n., "Le cyclone de Tamatave," La Nature, 1888, pp. 317-318 ; Le Temps, April 1, 1888.

[197] Set of documents in Archives Nationales de Madagascar, Antananarivo, Series DD.17; Correspondence concerning affairs in Madagascar, U.S. House of Representatives, Executive Documents Nos. 164 and 166 (50th Congress, 2nd Session); Lisagor, P. and Higgins, M., "Overtime in Heaven: Adventures in the Foreign Service," Garden City: Doubleday, 1964, Ch. 3, pp. 45-62; U.S. Department of State, State Magazine, Washington D.C., No. 414, May 1998, pp. 19-20.

[198] Sources: White Star Line ships at http://www.titanic-whitestarships.com/WSL_Asiatic.htm.

[199] Sources: Ship data sheet from the P&O Line (graciously provided by Ms. Susie Cox, Curator, P&O Heritage Collection); Clydebuiltships, Shipping Times, SS Peshawur, at http://www.clydesite.co.uk/clydebuilt/viewship.asp?id=15220

9

Russian Fleet 1904-1905[200]

RUSSIAN TRANSPORT SHIP (1905)

As the result of both Russia and Japan seeking to expand their empires into Manchuria and Korea, Japan opened hostilities with a surprise attack on the Russian naval base at Port Arthur (now Lüshun in China) in the night of February 8, 1904. This became the opening battle of the Russo-Japanese war. In reply to the resounding defeat of the Russian Far Eastern Fleet Tsar Nicholas II authorized a proposal from his government. Forty-seven ships from the Russian Baltic Fleet would sail 18,000 miles around the world from its bases in northern Europe, cross to the Far East, defeat the Japanese navy, and relieve Port Arthur. The epic journey that followed included an extended stay in the island of Nosy Be (northwestern Madagascar) that lasted from December 28, 1904 to March 16, 1905.

The Baltic Fleet, from which the Second Pacific Squadron[201] was drawn under the command of Vice-Admiral Zinovi P. Rozhdestvenskii, was a naval force in name only. This fleet including seven battleships led by the battleship *Kniaz Suvorov*, was a conglomerate of vessels with a wide range of design, builders, age and purpose. The craft rarely sailed, never exercised as a formation and rarely had gunnery exercises. Neither sailors nor officers had confidence in their ships. The four best battleships of the Boridino class were so top heavy that capsizing was always a concern. Machinery and gear were not maintained and repair was sloppy. The many different calibers of guns created a supply nightmare. The biggest problem was logistical. Ships of the period had to get new supplies of coal almost every two weeks. Unlike Britain and France, Russia did not have coaling stations scattered about the globe. As a result the admiralty contracted with an American company for the supply of the squadron with coal during the whole journey to the Far East. Re-coaling at sea was a dirty, dangerous business for the best of fleets not to mention the ill-trained Russian one.

The Russian admiralty compounded problems by insisting that virtually any ship that could float go on the journey. Rozhdestvenskii pointed out that these ships were primarily coastal, ill-armed, old, slow, and were coal guzzlers. He could only manage a compromise and had a number of useless, unseaworthy vessels foisted on him. Taking command formally in August 1904 the admiral instituted some training exercises but the results were very poor.

The Second Pacific Squadron sailed from Libav on October 15, 1904 with Rozhdestvenskii in *Kniaz Suvorov* leading forty-seven ships. In the region of Gibraltar the squadron separated. Its major forces under the command of Rozhdestvenskii finished the crossing going around Africa while a few ships whose build allowed them to use the Suez Canal went through the Mediterranean Sea to shorten the route. That detachment was under the command of Counter-Admiral D.G. Feljkerzam aboard the *Oslaybaya.* Another detachment headed by Captain L. F. Dobrotvorskii, left Russia when the Second Pacific Squadron was already on the way and also took the route through the Mediterranean Sea to the Suez Canal.

It was originally intended that the meeting place for the detachments would be Diego Suarez in the north of Madagascar. Rozhdestvenskii hoped to use the military port there to perform repairs. However, the French government, which was friendly towards Russia but was keeping neutral in the Russian-Japanese War, was swayed by the energetic diplomatic protests of Japan and Britain. It offered the Russian Squadron the bay of Nosy Be for their stay.[202] This enabled repairs to be done using only the squadron crew. However, the attitude of the French colonial administration and the French colony at Nosy Be towards the command of the squadron on an official as well as personal level was most friendly.

On December 28, 1904 the detachment of Counter-Admiral Feljkerzam, consisting of 2 battleships, 3 cruisers, 7 torpedo-boats and 9 transporters arrived in Nosy Be and was met by the French authorities. Twelve days later, on January 9, 1905 the major forces of the squadron arrived there lead by Rozhdestvenskii on the *Kniaz Suvorov* (6 battleships, 4 cruisers, 5 transporters and the hospital vessel *Orel*). On February 14 the squadron was joined by the detachment of Captain Dobrotvorskii, which comprised of 2 cruisers, 2 support cruisers and 2 torpedo-boats.

During the stay at Nosy Be several repairs, needed after the long ocean crossing, were performed on the ships and coal was loaded. Even so, news of the fall of Port Arthur on January 2, 1905 and the subsequent destruction of the First Pacific Squadron disheartened the rank-and-file as did news of revolts back home.[203] Rozhdestvenskii maintained discipline by constant training while at anchor in the bay and by sending the squadron out to sea several times to practice artillery shootings and to train on military maneuvers. To avoid being struck by Japanese ships reportedly cruising the Indian Ocean, patrols and security were organized at the entrance of the bay. During the night the ships were fitted with anti-torpedo nets.

Discharge ashore was infrequent and given only to officers and guards. Among the seamen, only the ill were discharged. In the morning hours they were allowed to go ashore in small groups accompanied by a doctor. The conditions of the stay, given the heat and high humidity were difficult. Alcoholism, illness, and clinical depression were rampant throughout the fleet. Because of health problems, 5 officers and 14 seamen were dismissed from the squadron. Also, a series of incidents took place during the stay resulting in the loss of eight crew members.[204]

The stay in Madagascar, which was supposed to last only a couple of weeks, turned out much longer than expected. The main reasons for the delay were twofold. First, the American coal supplier reneged on its contract because of a threat from Japan. Negotiations with the company continued till February when it finally agreed to continue to send coal shipments to supply the squadron. Second, during this period it was decided in St. Petersburg that the loss of the First Pacific Squadron would be made up by a Third Pacific Squadron under the command of Counter-Admiral Nicholas I. Nebogatov and that the Second Squadron had to wait in Madagascar for the reinforcements to arrive.

Rozhdestvenskii was staggered when receiving these orders. He strenuously objected to this decision stating that the long stay undermined the strength and morale of the Second Squadron while the Third Squadron made up of the same "self-sinkers" he had excluded from his fleet would not represent a serious military force. Furthermore, the delay in the start of the Second Squadron was giving the Japanese an opportunity to better prepare to meet it. Eventually, his arguments prevailed and it was decided that the Third Squadron would catch up with the Second Squadron at the coast of Southeast Asia.

After three months in Madagascar, the Second Squadron left Nosy Be. They sailed east on March 16, 1905 hoping to avoid the unwanted reinforcements. It left behind a very leaky transport ship to sink at anchor in Russians' Bay.

Rozhdestvenskii had to challenge the Japanese fleet commanded by admiral Togo with ships that could barely sail and crews that had hardly fired a live round. Notwithstanding tenacious determination, the Russian fleet was soundly defeated at the battle of Tsushima on May 27, 1905 obliterating Russian sea power. Both Japan and Russia accepted American President Theodore Roosevelt's offer of mediation on June 7, 1905. The ensuing Treaty of Portsmouth gave Japan economic and administrative rights in Korea and Manchuria. Russia's only territorial gain was half of the rugged Sakhalin Island.

REFERENCES

[200] Sources: Machikin, E., "Report on the Stay of Second Pacific Squadron in the Bay of Nosy-Be in Madagascar in 1904-1905," Research Historical Group of the Naval Fleet, Moscow, 2004; Ranurusehenu, H., "The Stay of the Russian Fleet in Madagascar in 1904-1905," Études Océan Indien, 18, 1994; Piouffre, G., "La guerre Russo-Japonaise sur mer," Nantes: Marine éditions, 1999, pp. 234-245; Cobb, J., "Russo-Japanese War, 1904-1905: A Naval Perspective (Part 2)," at http://www.gamesquad.com/distantguns/2006/02/21/russo-japanese-war-1904-%E2%80%93-1905-a-naval-perspective-%E2%80%93-part-2/; The Russo-Japanese War Research Society, at http://www.russojapanesewar.com/naval_links.html.

[201] On April 30, 1904, the ships at Vladivostok and Port Arthur were renamed the First Pacific Squadron. Ships drawn from the Baltic fleet were titled the Second Pacific Squadron.

[202] The ships were most of the time at anchor at the entrance of the bay of Passandava, a little south of Nosy Be, now called "Russians' Bay" ("Baie des Russes").

Figure 93. Monument erected in memory of the Russian sailors who died in Nosy Be

Figure 94. Monument erected in memory of the Russian sailors who died in Nosy Be

Figure 95. Monument erected in memory of the Russian sailors who died in Nosy Be

Figure 96. Monument erected in memory of the Russian sailors who died in Nosy Be

[203] Including the notorious "Red Sunday" (January 22, 1905) during which a large crowd led by the priest Gapone, while marching towards the Winter Palace to submit a petition to the Tsar, was fired upon by the police and the army, resulting in several hundred victims.

[204] On January 11, 1905, during the coal loading on the battleship *Borodino*, two engine mechanics suffocated in a side corridor. On January 12, 1905, on the support cruiser *Ural*, ensign Popov was killed by a falling arrow. On January 26, during maneuvers on the cruiser *Admiral Nahimov*, a marine fell overboard and drowned. On February 6, a boat was knocked down from the torpedo-boat *Blestiashtii (The Shining)* by a flying squall and 3 sailors drowned. On February 9, a sailor, who died on the cruiser *Dmitrii Donskoi,* was buried at sea.

10

Compagnie des Messageries Maritimes[205]

The origin of the Compagnie des Messageries Maritimes dates back to 1851 when Mr. Albert Rostand, a shipper in Marseilles, proposed a joint venture to Mr. Ernest Simons, the director of a surface shipping agency, to set up a maritime shipping line. First called Messageries Nationales, then Messageries Impériales, the line became the Compagnie des Messageries Maritimes in 1871. The Compagnie bought the naval wharfs of La Ciotat near Marseilles where most of its ships were built. The fleet was running both commercial routes and the postal service. The latter was subsidized by the French state. The company initially served the Levant and next South America. The Compagnie des Messageries Maritimes' golden age spanned 1871-1914. This coincided with French colonial expansion around the Mediterranean, in Asia, the Pacific and the Indian Ocean including Madagascar. Six of its ships foundered in Madagascar mostly due to being caught in cyclones.

Figure 97. Office of the Messageries Maritimes in Tamatave, early 20th century (Photography by P. Ghigiasso).

LE TAGE (1890)

The *Tage* was launched at La Ciotat for the Compagnie des Messageries Maritimes on May 3, 1868 as the first of three sister ships. Originally at 1,691 tons, she was driven by a two-cylinder pounder steam engine. She originally plied the Levant route. In 1888, she was lengthened, refitted and her tonnage increased to 1,854 tons with the intent of being stationed in Madagascar and used for local navigation which she undertook from 1889 onward. On January 14, 1890 the *Tage* was taken in a storm and wrecked on Nosi Mahampana (Barracouta Islands, close to Vohemar, along Madagascar's northeast coast) and now rests in about 10 meters of water.

DOURO (1910)

The *Douro*, a 2,700-ton mixed passenger-cargo vessel driven by a single triple expansion steam engine heated by two coal boilers was launched at La Ciotat on February 16, 1889. Initially employed on the London-Le Havre-Marseille route she took part as a troop transport vessel to Madagascar for the French expedition of 1895. In 1903 she ran the Mediterranean and Black Sea route followed by the Indochina route from 1908 onward. Next she was assigned on the Madagascar route. She ran aground and was lost in a storm in Farafangana in the southeast of Madagascar on May 12, 1910.

SALAZIE[206] (1912)

The *Salazie*, a 4,256-ton passenger ship powered by a compound 3-cylinder steam engine heated by eight coal boilers (and a single tri-

Figure 98. Wreck of the *Tage*. Steering gear. (Reproduced with the kind permission of Eric Gilli)

Figure 99. Wreck of the *Tage*. An anchor. (Reproduced with the kind permission of Eric Gilli)

Figure 100. Wreck of the *Tage*. Propeller (Reproduced with the kind permission of Eric Gilli)

Figure 101. The *Douro* in Marseille (Reproduced with the kind permission of Mr. Philippe Ramona)

Figure 102. The *Douro* in Marseille (later picture, the yard is gone) (Reproduced with the kind permission of Philippe Ramona)

ple expansion engine from 1895 onward) was the fifth of a series of seven identical vessels. She was launched in La Ciotat on April 8, 1883 for the Compagnie des Messageries Maritimes. From 1883 through 1890 she plied the Marseille-Nouméa line via Reunion and

Figure 103. The *Douro* leaving Marseille (Reproduced with the kind permission of Mr. Philippe Ramona)

Sydney. Starting in 1891, the *Salazie* sailed the runs to China. After 1904 she sailed the lines to Egypt, the Far East or Madagascar according to needs.

On her last fateful voyage, under the command of Captain Albéric Laréquier, the *Salazie* made several calls before crossing the Equator and then called on Zanzibar, the Comoros, Mahajanga and Nosy Be.

Figure 104. The *Salazie* in Marseille (1895-1905) (Reproduced with the kind permission of Mr. Philippe Ramona)

Figure 105. The *Salazie* repainted in black (1905-1912) (Reproduced with the kind permission of Mr. Philippe Ramona)

Figure 106. *Salazie* (1905-1912) (Reproduced with the kind permission of Mr. Philippe Ramona)

Figure 107. The *Salazie* wrecked at Nosy Akoumby (Reproduced with the kind permission of Mr. Philippe Ramona)

She anchored in Diego Suarez on November 23, 1912 where military and naval officers and their families disembarked. Many passengers went ashore and visited Antsiranana during the day. After re-coaling the *Salazie* weighed anchor at 9:30 pm, bound for Tamatave, Reunion and Mauritius. The heat was oppressive and the sea calm during the afternoon but the wind picked up and the sea swelled noticeably in the evening. During the night, on a southward course along Madagascar's east coast, the wind became ferocious and enormous waves hurled the ship. The Captain grumbled that he had not been told about the strength of this cyclone or else he would have stayed in Diego bay.[207]

The long night passed but the morning of November 24 was even more ominous. The sky was pitch-dark and the ship was trembling under the force of waves, up to 120 feet high hitting mainly on starboard so she persistently leaned to the port side. The crew started closing most hatches but were a bit late. Large amounts of water ran into the passageways. By eleven in the morning the upper deck had suffered major damage, all deck chairs were gone, the grand piano had crashed and the kitchen was flooded so the cook apologized that only a cold lunch would be served. Hardly any passengers had lunch anyway. In the afternoon the damage worsened. Several of the boats, including the long boat, swung from starboard to port, breaking the fasteners and were thrown into the ocean; furniture and decorations were swinging widely and shattered; waves on the starboard side crushed some of the doors, inundated the bridges and progressively flooded the larboard cabins.

All the pumps were used to protect the engine which was working at slow pace since seven in the morning. At three in the afternoon the engine stopped completely because of insufficient pressure and because ropes floating with other debris wound around the propeller and immobilized it. The Captain realized that his ship was unsteerable. The first mate on the port side threw a flood anchor to attempt to haul the ship against the wind but a colossal wave threw him in the water and he drowned. Both the mizzen mast and the main mast crashed with a thunder and the railing became torn in several places. The electricity stopped completely when the batteries flooded. By nightfall the *Salazie* was pitching more and more on the port side. The passengers, strapped in their life jackets, prayed and held hands, awaiting a certain death on this floating coffin drifting aimlessly.

Around 8:30 pm two huge jolts a few minutes apart shook the ship. Menacing cracks made the passengers believe that they were sinking. The dreadful shocks caused the engine to shift bursting two steam pipes in the process. Acrid steam spread everywhere, asphyxiating passengers and forcing them to run or crawl for the bridges. Unbeknownst to the passengers an enormous wave had thrown the *Salazie* on the rocks of Nosy Akoumby north of Vohemar. A few minutes later another huge wave lifted the ship anew and thrust her even farther on the rocks. The pounding surf would have broken the *Salazie* but, miraculously, the wind started dropping and the barometer shot up as by magic.

The steam dissipated progressively and the droplets of boiling water ceased to burn the passengers. Shortly thereafter the Captain entered the main cabin. He announced that the ship had run aground on a small islet and that everybody was alive except for the first mate who drowned earlier. For the passengers it was a true awakening from an endless nightmare and they were overjoyed. The night was nevertheless painful, the ship still battered by the remnants of the storm and the exhausted passengers drenched. In the early morning of the 25th the crew and passengers disembarked on Nosy Akoumby in a battering rain. Tents were raised and a chain was set up to carry food and luggage ashore. Three days later two ships came to the rescue and brought the crew and passengers back to Diego Suarez.

Figure 108. The *Salazie* wrecked at Nosy Akoumby with a tent in the foreground (photo Alain de Bressy)

IMERINA (1932)

The *Imerina,* a 2,830-ton passenger ship powered by an alternative triple expansion engine, was launched in Stockton in 1899 for the German company Freitas under the name of *Sparta*. She was sold in 1900 to the Hamburg America Line. In 1914, the Compagnie des Messageries Maritimes purchased her for local navigation around Madagascar and renamed her *Imerina*. She plied the Madagascar coast for close to 20 years. On December 1, 1932 under Captain Drevet, she hit the reef off the entrance of the harbor of Manakara, ran aground on the beach and was abandoned.

BAGDAD[208] (1935)

The *Bagdad*, a 2,832-ton mixed passenger-cargo ship powered by an alternative triple expansion engine heated by two coal boilers, was launched in 1891 in Dundee under the name *Athenai*. She was purchased from Panhellenic S.S.Co, Piraeus in 1895 renamed *Bagdad* and used for Madagascar coastal service. She survived a number of cyclones and incidents. On January 27, 1913 she ran aground at

Figure 109. The *Imerina* (Copyright Marius Bar—Toulon (France))

Ambilobe due to shifts in sandbanks caused by the swollen river. On October 19, 1917 she hit the reef south of the Golo Pass at Tintingue; she was saved but needed 15 days of repairs at Tintingue before being able to proceed to another port for more extensive repairs. In 1923 her fittings were destroyed by fire at Fort Dauphin but she was able to make it to Diego Suarez for repairs. In 1924, then under Captain Preneyre, the *Bagdad* touched bottom while attempting the inner anchorage at the harbor of Manakara; she had to be fully unloaded to get out of this predicament. After serving nearly 40 years in Madagascar on November 29, 1935, under Captain Gilbert, the *Bagdad* wrecked on cape Lahatrazona in the northeast of Madagascar. There were no casualties.

AMIRAL PIERRE (1942)

The *Amiral Pierre,* a 4,391-ton mixed passenger-cargo vessel, was built in 1905 by Barclay Curle & Company in Glasgow. Launched as *Newby Hall* for the Ellerman Lines Ltd, she was sold in 1930 to the Greek shipper G. Andreou and renamed *Yiannis.* She was purchased in 1940 by the Compagnie des Messageries Maritimes and renamed *Amiral Pierre.* In 1941 she was acquired by the French Vichy government in Madagascar. She was scuttled and sunk in the Mozambique Channel off the south of Madagascar[209] on September 30, 1942 when intercepted by the British destroyer *HMS Nizam.*

Figure 110. *Bagdad* in the harbor of Marseille (reproduced with kind permission of X. Escaillier)

REFERENCES

[205] Sources: Commandant Lanfant, "Historique de la Flotte des Messageries maritimes-1851-1975," Association des Anciens des Etats-majors des Messageries Maritimes, second ed., 2001; Bois, P., "Le grand siècle des Messageries Maritimes," Chambre de Commerce et d'Industrie de Marseille Provence, 2ème édition, 1992; Web site of the Messageries Maritimes at http://www.messageries-maritimes.org/.

[206] Sources: Website of the Messageries Maritimes at http://www.messageries-maritimes.org/; Campiniano-Cantemir, J., « La Fin du Salazie, par un Naufragé, » Paris: Chamolle, 1913.

[207] This is debatable: the inquiry later demonstrated that the wreckage was largely due to the foolhardiness of the Captain, who believed he could pass before being hit by the cyclone.

[208] Sources: Web site of the Messageries Maritimes at http://www.messageries-maritimes.org/; Lhuillier, M., "Naufrages, Echouements et Accidents de Navigation survenus à Madagascar," Bulletin de l'Académie Malgache, T. XXIX (1949-1950), p.86-88.

[209] At 26°04 south and 34°54 east (source: Saibène, M., Brouard, J.Y., and Mercier, G., "La Marine Marchande Française 1940/1942," Vol. II).

11

Compagnie Havraise Péninsulaire

The Compangnie Havraise Péninsulaire (CHP) is one of the main shipping lines continuously serving the Indian Ocean route by way of mixed passenger-cargo vessels since the late 19th century.[210] The CHP was created in 1882 as a joint venture between two co-founders, Messrs. Hentsch and Mannberguer and Mr. Grosos, a shipper from Le Havre who operated since 1865. In the early years, the CHP served the ports of Spain, Algeria and Marseille from Le Havre. Development was rapid. By 1885 CHP ships were plying the Antilles, Vietnam, the Persian Gulf, the Mozambique Channel and the Indian Ocean. In 1886, Madagascar was added with ships leaving every 45 days and every month from 1891 onward. The French conquest of Madagascar under General Gallieni in 1895 opened new prospects. There was a service to the Indian Ocean and Madagascar every 15 days with service to secondary ports assured since 1897 by small coastal vessels, initially the *Tafna*.[211] Seven of its ships foundered in Madagascar; the last under Malagasy flag.

VILLE DE RIPOSTO[212] (1899)

The *Ville de Riposto,* a 1,494-ton three masted cargo schooner powered by a compound steam engine, was built in 1884 at the R.

Thompson yards in Sunderland, U.K. for the CHP. In her early life she plied the Spain-Mediterranean route out of Le Havre. In 1892 off Gibraltar, she saved the crew of the sinking French schooner *Edouard.* In September 1898, the *Ville de Riposto* was sold to the Société Française de Commerce et de Navigation of Madagascar and used for coastal service. Having left Tamatave and sailing on a southern course with a crew of 33 and nine passengers and a cargo of rice, salt and other wares, she hit an unknown reef off Farafangana (southeast coast) during nice weather on January 29, 1899 and was lost. There were no casualties.

VILLE D'ALGER II[213] (1920)

The demise of the *Ville d'Alger II* on February 1, 1920 was by far the worst catastrophe that befell the CHP. She was a 4,857-ton two-masted steamer with three-decks powered by a triple expansion engine. She was built in 1912 by the Société des Forges et Chantiers de la Méditerranée at Graville and used since 1912 on the Le Havre to the Indian Ocean route.

At 9 that morning of February 1, 1920, under the command of Captain Rebours, the *Ville d'Alger II* weighed at Pointe-des-Galets (Reunion) bound for Europe with 91 passengers and 51 crew members aboard and a cargo of rum and sugar. At 8 that evening, her position was 20°25' south and 50° east opposite Madagascar's east coast 7° (or 420 nautical miles) south of Tamatave. She sent a distress signal of a fire having broken out aboard. The signal was heard by several ships moored in Tamatave that immediately came to the rescue.

The mixed CHP passenger-cargo *Ville du Havre* weighed at 10 the next morning and three days later found the abandoned and still burning wreck drifting further out in the ocean (18°54' south and 51°58' east). The Greek steamer *Parnassos* and the American cargo ship *Westport* likewise scouted the disaster area but in vain. More than a week later, on February 9, a boat landed close to Foulpointe. She had 21 people on board including the second-in-command; the only survivors of this tragedy. They testified that all the passengers and crew had boarded the boats but that all except theirs must have capsized due to the heavy seas. They were brought back to Tamatave aboard the *Persepolis,* a coaster used by the Messageries Maritimes. A little later, the wreck was found stuck to the reef of one of the small Barracouta islands in the northeast.

CATINAT[214] (1927)

One of the most violent cyclones ever hitting Tamatave was on March 3, 1927. Since February 24, the weather service in Antananarivo was anxiously following the development of a massive cyclone in the Indian Ocean. It followed a north-south course well east of Madagascar, threatening first Mauritius then Reunion Island. Then a unique phenomenon happened: the cyclone turned around from its southern route to take on a northwestern course and headed from Reunion straight to Tamatave. Southwestern winds blew at more than 125 mph. A tidal wave measuring 2.8 meters washed over the city killing about 600 people and transforming the city in a jumble of ruins. The 300-meter long wharf built by Levallois-Perret in 1900 was completely destroyed. In 1929, this resulted in the construction of new docks that are still in operation today.

The cyclone created havoc among the vessels anchored in its harbor. Two steamers from the CHP, the *Catinat* and the *Ville de Marseille II* ran aground. The *Ville de Marseille II* was a 5,071-ton mixed passenger-cargo steamer launched in 1910 and in service since 1911 plying the Indian Ocean route. The *Catinat,* a 7,544-ton mixed passenger-cargo steamer powered by a triple expansion engine, was built in 1913 at the Tecklenborg yard at Bremerhaven (Germany) for DDG Kosmos in Hamburg and named *Menes*. In 1919 she was handed over to the French government as part of war damages and attributed to the Société Générale d'Armement which named her *Catinat*. She was purchased by the CHP in 1924 and ran the Le Havre-Marseille-Algeria-Indian Ocean route.

On March 3, 1927, both ships were anchored in the harbor of Tamatave. Before sunrise the barometric pressure dropped to very low levels, the southwestern gale blew ferociously and a heavy swell unfurling in the pass made any attempt to leave the harbor impossible. By seven that morning the swell was unfurling throughout the harbor. Both ships were running their engines at full steam forward to lessen the exertion on the chains and avoid being thrown on the rocks of Tanio point.

Around 8:30 am, the *Ville de Marseille II* dragged her anchors notwithstanding the full steam forward. Sliding backwards, she avoided Tanio point but was thrust against the great reef on her port side. The southwestern crosswind glued her against the reef. A brutal surf pounding her, the engines had to be shut down. Around 10:15 am, the much heavier *Catinat* was also heaved backwards after ripping her two anchors and ran into the *Ville de Marseille II*'s starboard side causing significant damage. The *Catinat* then glided alongside the *Ville de Marseille II* that was sandwiched between her and the reef, taking away the starboard boat and causing further damage. The *Catinat* kept being heaved backwards and eventually ran aground at the mouth of the river Ivoloina 7.5 miles north of Tamatave. She broke into two at the back of the funnel. She was sold in June of that year and torn down in 1945.

By noon the *Ville de Marseille II* commanded by Captain Forgeard and still glued to the reef, found herself in the eye of the cyclone. Taking advantage of the temporary lull she bore east to leave the breakers and sailed on a southeastern course. For a second time she got hit by the cyclone that sent her back on a northern path. By late afternoon, when the winds finally subsided, she was back opposite Tamatave and managed to anchor in the harbor in the evening. Due to the damage sustained during the collision with the *Catinat* she started taking in much water in the starboard holds and the listing became gradually worse. During the whole of the next day, with the pumps failing at times, the crew fought a loosing battle with the rising water and listing.

Figure 111. The *Catinat* after running aground at the mouth of the Ivoloina river (reproduced with the kind permission of Charles Limonier)

Early in the morning on the 5th the water intake worsened and started to invade the boilers. The Captain tried to find additional external pumps but Tamatave was a heap of ruins and nothing could be had. At 8:30 am, the listing worsened and starboard boiler flooded. The engines would shut down soon. Faced with the severity of the situation, Captain Forgeard decided to beach the *Ville de Marseille II*. He weighed and cast off, beaching the ship around 9 am in the harbor on the north side of the wharf. When beached, the Captain anchored the port side and placed a hawser on the trunk buoy to prevent the ship to fall off. In the afternoon, the hawser snapped and pushed by the surf and a strong southeastern wind, the ship was pushed further ashore. Twelve days later the *Ville de Marseille II* was put back afloat. She sailed to Marseille for repairs but suffered a major fire in one of the holds and was beached near the harbor. She eventually made it to Le Havre and was sold in 1935 to a Belgian demolition wharf and towed to Antwerp.

VILLE DE MAJUNGA[215] (1928)

The *Ville de Majunga* was a 6,500-ton mixed passenger-cargo two-masted steamer with three-decks and a spar deck. She was built by

Figure 112. The *Ville de Marseille II* beached at Tamatave (reproduced with the kind permission of Charles Limonier)

the shipyard Sir James Laing & Sons of Sunderland (UK), launched in November 1900, and entered service for the CHP on February 28, 1901. She was part of a series of eight similar ships of which the *Ville de Tamatave* was the flagship. She was powered by a triple expansion engine with two boilers and a single steel propeller developing a cruising speed of 10 knots. Her crew numbered 45 and she was equipped for 20 passengers and the bulk transport of wine on the Madagascar run.

On March 17, 1928, coming from Reunion under the command of Captain Masson, she touched a shoal in the harbor of Mananjary (southeast coast). Though she sprung an important leak and started to take in much water, Captain Masson tried to sail her to Diego Suarez for repairs. Unfortunately, the pumps could not handle the water intake and soon all the holds were filling up rapidly. The decision was taken to run the ship aground. The crew was saved but the *Ville de Majunga* remained beached at Mahanoro. Efforts to put her back afloat failed.

VILLE DE DJIBOUTI[216] (1928)

The 4,376-ton mixed passenger-cargo steamer *Ville de Djibouti*, powered by a three-cylinder compound steam engine, was built in

Figure 113. The *Ville de Majunga* (reproduced with the kind permission of Charles Limonier)

1914 at the Russell yards in Glasgow (U.K.) for the Gow Harrison & Co. of Glasgow and originally named *Verdun*. In January 1915 she was purchased by the Bay Steamship Co., charged by the French government during the First World War to buy ships to be run under British flag, and renamed *Bay Verdun*. In 1921, when the British government acquiesced to the flag transfer, she became part of the French government's fleet and renamed *Port du Havre*. Upon liquidation of the French government's fleet in February 1922 she was sold to the CHP which renamed her *Ville de Djibouti*. She was then used on the Le Havre to the Indian Ocean route via the Mediterranean. The *Ville de Djibouti* ran aground in a violent cyclone on April 19, 1928 three miles south of Mananjary and became a total loss.

VILLE DE PARIS II[217] (1935)

The 5,020-ton mixed passenger-cargo two-masted steamer with three-decks and a spar deck *Ville de Paris II* was built in 1903 for the CHP in Dunkirk. She was powered by a triple expansion engine with two boilers and a single steel propeller developing a cruising speed of 10.5 knots. She housed 22 first-class, 30 second-class passengers and a crew of 45. After being leased in 1904 for three years to Lamport &

Figure 114. *The Ville de Paris II* (reproduced with the kind permission of Charles Limonier)

Holt to sail the west coast of South America she served the Indian Ocean line during the rest of her life.

On February 16, 1935, under the command of Captain L'Herrec, the *Ville de Paris II* left Le Havre bound for Reunion via Pauillac, Algeria and Marseille. Next she called on Fort Dauphin. On May 8, 1935, while sailing from Fort Dauphin to Farafangana, she hit a sandbank in full daylight 6 miles off the coast at Saint-Luce most likely due a navigation error. The ship sank in a few hours. There were no casualties and the crew returned to France on the *Bernardin de Saint-Pierre* of the Messageries Maritimes.

VILLE DE MANAKARA III[218] (1994)

The *Ville de Manakara III*, a 1,400-ton coastal ship with two decks and bridge at the back, powered by a diesel engine, was built by the Bijker's Aannemings Bedrijf N/V yards of Gorinchem (Holland) in 1960 and entered service in 1961 for the Danish shipping company Ole Danielsen under the name *Ulla Danielsen*. She was purchased by the CHP in 1964 and renamed *Ville de Manakara*. She immediately sailed to the Indian Ocean where she was used for local navigation. In November 1964, she was freighted by the CHP to the Compagnie Malgache de Navigation (CMN). The transfer under a Malagasy flag took place on December 1, 1964.

In November 1969, the *Ville de Manakara III* was sold to the CMN. She continued to ply local waters and regularly called at Comoros, Reunion and Mauritius. She sank in the harbor of Diego Suarez in 1994. Her wreck will be taken apart in 2009 as part of a program to remove wrecks that get in the way of navigation channels undertaken by the Agence Portuaire Maritime et Fluviale de Madagascar.

REFERENCES

[210] The main source used for the introductory paragraphs is the excellent compendium of the CHP by Limonier, C., « Les 110 Ans de la Havraise Péninsulaire, Histoire de la Flotte », Marseille: P. Tacussel, 1992.

[211] The CHP still has minority ownerships in two affiliated companies in Madagascar, the Société Malgache de Transport Maritime and the Société Auxilière Maritime de Madagascar (Auximad), both of which are now under liquidation.

[212] Sources: Limonier, C., op. cit., pp. 84-85 ; Valette J. and Ratsimbazafy, A., "Les naufrages et échouements survenus à Madagascar de 1898 à

Figure 115. The *Ville de Manakara III* (reproduced with the kind permission of Charles Limonier)

1901," Bulletin de Madagascar, No. 235, December 1965, p. 1057; Miramar ship index, at http://www.miramarshipindex.org.nz/ship/list.

[213] Limonier, C., op. cit., pp. 140-141.

[214] Sources: "Histoire de Tamatave" at http://tamatave.ifrance.com/histoire/histoire.htm; Limonier, C., op. cit., pp. 125-131 and 181-182; and Chauvin J., op. cit, pp. 162-167.

[215] Sources : Limonier, C., op. cit., pp. 109-110 ; and s.n., « Perte du Ville de Majunga » at http://naviresdeguerre.free.fr/majunga.html

[216] Sources: Limonier, C., op. cit., pp. 171-172; and Miramar ship index, op. cit.

[217] Source: Limonier, C., op. cit., pp. 115-116; Miramar ship index, op. cit.

[218] Sources: Limonier, C., op. cit., pp. 277-278.

12

The Battle of Diego Suarez[219]

BOUGAINVILLE, HMS AURICULA, BÉVEZIERS, HÉROS, AND MONGE (1942)

The French naval base at Antsiranana (Diego Suarez) in the North of Madagascar, completed in 1935, was France's most modern colonial port with a dry dock that could accommodate 28,000 ton battleships and an arsenal capable of repairing the largest guns. After the Second World War broke out Madagascar acquired an immense strategic importance. At that time, the vital convoys supplying and reinforcing the British forces in the Middle East and India rounded the Cape of Good Hope and then sailed through the Mozambique Channel. Clearly, whoever held Diego Suarez controlled the western Indian Ocean. As the French authorities in Madagascar were firm supporters of the German-influenced Vichy government, Allied leaders feared that ports on Madagascar might be used by Japanese naval forces and threaten the Allied war effort and merchant shipping in the Indian Ocean.

Although between December 1941 and January 1942 Axis negotiations set longitude 70 degrees east as the boundary between German and Japanese naval operations in the Indian Ocean exceptions were to

Figure 116. Diego Suarez' dry dock today

be allowed as circumstances warranted. In the evening of March 5, 1942 in the Führer-Hauptquartier in Berlin, Grand Admiral Erich Raeder briefed Adolf Hitler and his staff—not altogether accurately—that based on reports he had seen, the Japanese were planning to establish bases on Madagascar and Ceylon (now Sri Lanka) to control sea traffic in the Indian Ocean and Arabian Sea. Raeder mentioned that Germany should consent to these Japanese plans. For Madagascar, the Japanese would require approval from their German allies (for it lay on the German side of the boundary line) and from the Vichy regime (who controlled and defended Madagascar). Hitler, seldom interested in naval affairs, was unenthusiastic and replied that he doubted that the Vichy French would consent to Japan establishing bases on the island.

Despite Hitler's lack of interest, Churchill needed a symbolic victory to boost the morale of his troops and British public opinion. In Asia, his naval dissuasion forces were obliterated in December 1941, and Hong-Kong, Singapore, Rangoon, Java and Borneo had fallen in the hands of the Japanese in the first quarter of 1942. The aero naval battle in the eastern Indian Ocean in early April 1942 had resulted in a Japanese victory with heavy casualties on the British side. In Africa, the Italo-German counter-offensive in Libya led to the fall of Benghazi in January 1942. In early May, Rommel was preparing his offensive against El Gazala and Tobruk and Egypt was next. Finally, German U-Boats wreaked havoc on British and American naval convoys. Madagascar, lightly defended and politically fragile, gave him the opportunity for a quick victory with two other benefits: First, it would prevent the Japanese from entering this strategic sector. Second, it would offer an excellent training ground for coordinated Navy, Fleet Air Arm and land forces.

Churchill telegraphed Roosevelt: "A Japanese air, submarine and/or cruiser base at Diego Suarez would paralyze our whole convoy route both to the Middle East and to the Far East." Field Marshal Smuts cabled Churchill that Madagascar is "the key to the safety of the Indian Ocean" and feared that the Japanese might use bases on the island in an advance against the African mainland in the same manner that they had recently used bases in Indo-China in their advance against Burma, Malaya, Singapore, British Borneo and the Netherlands East Indies. The official motive for the operation was thus sealed. De Gaulle tried to object and wished to make a Free French landing on Madagascar but his failure at Dakar in September 1940 meant that his plans found no support.

In the spring of 1942, it was decided to launch an amphibious assault to take Diego Suarez; called "Operation Ironclad." Under the command of Rear-Admiral Syfret, a large force of ships assembled at Durban, South Africa towards the end of April 1942 awaiting the War Cabinet's final approval of the invasion. The orders given; a force of some 13,000 troops with tanks and artillery, supported by 46 warships and transport vessels and 101 aircraft assembled to the north of Amber Cape before dawn on May 5. The Royal Navy ships included the battleship *HMS Ramillies,* aircraft carriers *Indomitable* and *Illustrious,* cruisers *Hermione* and *Devonshire,* eleven destroyers, numerous frigates, corvettes and troop transports. The narrow entrance to Diego Suarez Bay was known to be powerfully defended by large-caliber artillery so it was decided to land in Courrier and Ambararata Bays on the northwest corner and march across country to take Antsiranana from the landward side.

The assault forces first captured a small battery overlooking Courrier Bay. With the beaches secured, the main forces landed under cover of the Royal Navy ships and began their march towards Antsiranana without encountering major resistance at first. However, by mid-morning, the advance was held up on the outskirts of the town. Some three miles to the south of Antsiranana, General Joffre had built a strong defensive line across the isthmus of the peninsula in 1909. It was comprised of a trench network and an anti-tank ditch strengthened by forts and pillboxes housing artillery and machine-guns. For two days, the invading forces assaulted the French line however without success and with fairly heavy casualties.

Meanwhile, the Fleet Air Arm bombarded the airfield obliterating the small French air detachment and depth-charged and torpedoed the anchored French fleet. They first zoomed in on the largest ship, the colonial aviso *d'Entrecasteaux,* which could leave her berth and escape. She took up a strategic position in Diego Bay north of Cape Diego to provide supporting fire against the northward column of the invading land forces. The British planes attacked and sunk the cruiser *Bougainville,* the submarine *Beveziers* and two Italian ships, the *Somalia* and the *Duca degli Abbruzzi.*

The *Bougainville*[220] was originally called *Victor Schoelcher,* a 4,509-ton banana transport ship built at La Seyne-sur-Mer and launched on February 25, 1938 for the French Ministry of Marine Shipping. Originally leased to the Société Générale de Transports Maritimes à Vapeur and plying the Antilles line she was transferred to the Compagnie de Navigation Fraissinet in April 1939. She was requisitioned by the French Navy from September 28, 1939 through October 12, 1940 and armed as an auxiliary cruiser. In April 1941, the *Victor Schoelcher* went back to commercial shipping but was armed again in late 1941 and renamed *Bougainville.* After being assigned to Madagascar she participated in April 1942 with the *Beveziers* in supplying Djibouti. This forced the blockade set up by the Allied forces

Figure 117. The *Ramillies* (reproduced with the kind permission of Mick French on behalf of the HMS Ramillies Association—not to be reproduced without permission).

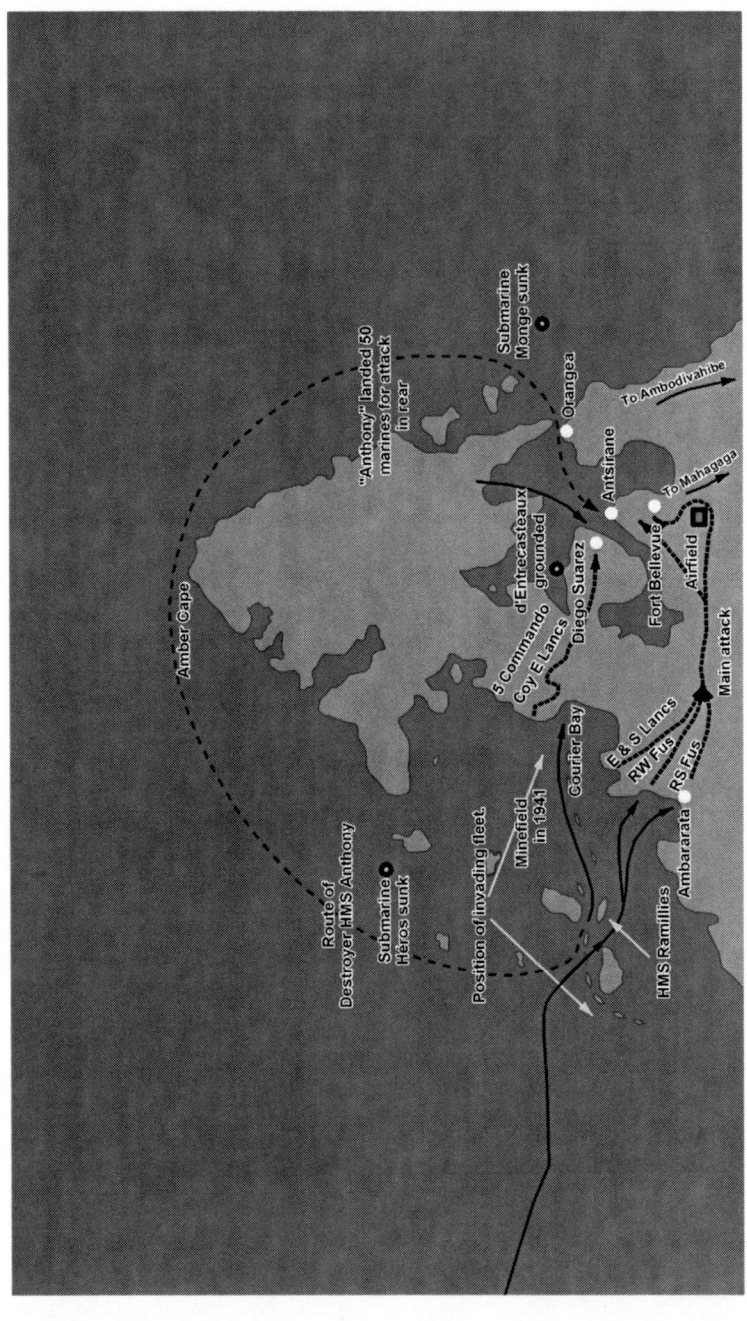

Figure 118. Operation Ironclad

since June 1941. Under the command of Capitaine de frégate Fontaine she became the first naval victim of Operation Ironclad being sunk shortly before 6:00 am on May 5, 1942 by torpedoes from aircraft of the *Illustrious*. The *Bougainville* still rests in the Bay of Diego Suarez.

The *Beveziers* submarine was launched on October 14, 1935.[221] Initially based in the Antilles, she joined Halifax and next the French West-African forces. She was in Dakar during the Franco-British assault of September 1940 and torpedoed *HMS Resolution* on September 25, 1940. Back in Toulon in January 1941 for a major careening the *Beveziers* was redeployed to Madagascar arriving in Diego Suarez on March 19, 1942. She left Diego on April 6 with the *Bougainville* to supply Djibouti forcing the Allied blockade. She returned to Diego on April 25. At 6 in the morning on May 5, 1942 then under the command of Lieutenant de vaisseau Richard, she was attacked and sunk while trying to leave her berth in the harbor of Diego Suarez by a flight of British Faireys *Swordfish* torpedo-bombers from the carrier *Illustrious*. She sunk in the harbor as her ballasts were punctured. Five sailors were killed in the attack but the rest of the crew escaped. *Beveziers* was put back afloat by the Allied forces in April 1943, put in reserve and finally taken out of service on December 26, 1946. She now rests, partially immersed in a few feet of water on an embankment at the south side of Diego Suarez' harbor.

Around noon on May 5 to the west of Courrier Bay, the British corvette *HMS Auricula* captained by Lt. Commander S.L.B. May-

Figure 119. The *Beveziers* in 1942 (photo courtesy of Bertrand Godin)

Figure 120. The *Beveziers'* resting place today

Figure 121. The *Beveziers* being scuttled (reproduced with the kind permission of Tim Healy)

bury, was tasked to clear a channel through a minefield ahead of the landing craft and struck a mine. This broke her back but she continued to float and remained anchored to her sweep. Several members of her ship's company were wounded but none was killed. Later the *Auricula* was taken in tow by the corvette *Freesia* but broke in two and sank in the morning of May 6 after her crew had transferred to other ships. The *Auricula* was a 925-ton a Flower Class corvette built in 1940 at W. Simons & Co. and powered by quadruple-expansion engines. She became famous for towing a Sunderland coastal command flying boat, which had suffered a complete loss of oil pressure on the starboard inner engine, for three days against a heavy swell and a fierce crosswind for about 350 miles to her base at Freetown, Sierra Leone in February 1942.[222]

To unlock the stalemate on the ground, the British command decided during the night of May 5-6 to concentrate forces for neutralizing the *d'Entrecasteaux* and to prepare a rearguard action to separate the French forces by dispatching a Royal Marine unit from the *Ramillies* to take the French defenses from the rear. The *d'Entrecasteaux* was pounded by Swordfishes of the *Indomitable* in the morning of May 6 which set her on fire and finally grounded her.[223] The marines were embarked in the evening of May 6 in the destroyer *Anthony* which rounded the northern tip of Madagascar through the night and succeeded in entering Diego Bay undetected. After landing in the harbor, the Marines came to a large barracks building and caught the French soldiers asleep with their firearms piled neatly in the entrance to their dormitories. The French surrendered and the Marines set to the town. This rearguard action broke the French resolve and when the main frontal attack was renewed later during the night the French line was overrun.

The first and sole air combat in Madagascar's skies took place that morning of May 7.[224] A squadron of three French MS406 planes, on a surveillance mission to the south of Diego Suarez, were intercepted by a flight of Martlet from the carrier *Illustrious*. One of the British combat planes was hit but managed to crash land on a beach while the three French planes were shot down. Diego Suarez surrendered in the afternoon of May 7 and the important anchorage was in British hands. The French forces withdrew to the south.

While the fighting on land was over in the afternoon of May 7, 1942 the naval battle lasted a bit longer. In the morning of the first day of Operation Ironclad, the French naval commander Maerten had ordered the three other French submarines operating in the area, the

Glorieux, the *Héros* and the *Monge* to immediately set sail for Diego Suarez and to torpedo any enemy vessel in sight.

The *Glorieux* was launched in November 1931 and commissioned in 1933 in the Gulf of Gascony. She participated with the *Héros* in the defense of Dakar in 1940 and underwent careening in Toulon before being redeployed to Madagascar arriving in Diego Suarez in late November 1941. She participated in forcing the Allied blockade to supply Djibouti. When commander Bazoche received Maerten's message the *Glorieux* was calling at Mahajanga. She immediately set sail on a northward course and dove in the morning of the 6th twenty miles south of Diego Suarez.

At 10 in the morning on the 6th, the *Glorieux* sighted the aircraft carrier *Indomitable* escorted by three destroyers sailing east but their speed prevented a torpedo launch. There were further sightings in the afternoon of the 6th and a few during the next day but too distant for a torpedo launch. The *Glorieux* resurfaced during the nights of the 6th and the 7th to recharge her batteries not having fired a single shot and without any news from Diego Suarez. During the 8th, she continued her search but in vain. The next night she received the news from the French Admiralty of the surrender of Diego Suarez. Commander Bazoche sailed south to Nosy Be. On the 11th, he was ordered to join Port-Androka and thence sailed back to Dakar. The *Glorieux* was scuttled in Toulon in November 1942 just before being captured by the Free French forces. She was put out of service in 1952.

The *Héros* was built in Brest launched in October 1932 and commissioned in the Atlantic in 1934. At the beginning of the war she operated off Morocco and the Canary Islands before joining Dakar in January 1940. The *Héros*, together with the *Glorieux*, returned to Toulon in November 1940. After being careened, she left, again with the *Glorieux* for Dakar in September 1941 and then proceeded to Diego Suarez arriving on November 27, 1941. On February 8, 1942 the *Héros* escorted the *Bougainville* and forced the blockade at Djibouti relieving the *Glorieux*. She returned to Diego Suarez on March 25, 1942.

When Maerten issued his orders, the *Héros* was located 500 miles to the north of Madagascar escorting a supply cargo bound for Djibouti. As soon as he received the message Commander Lemaire turned around, leaving the cargo in the Comoros, and headed south. On May 6 the *Héros* was attacked by a British plane 250 miles north of Madagascar but she could dive in time. Resurfacing 30 minutes later, she continued on her southward course. However, due to lack of

Figure 122. The Submarine *Le Héros*

communications with his base, commander Lemaire underestimated the British operation and the aerial attack that had just taken place believing that the plane had come from Seychelles and not from an aircraft carrier.

The *Héros* rounded Cape Amber at 3 in the morning on September 7. At 4:30 in the morning she arrived in Courrier Bay sailing at the surface. At 5 in the morning, a plane was spotted on the starboard side. It was part of a flight of Swordfish from *Illustrious* that had located the submarine in the dark on their radars. The *Héros* crash-dived to a depth of 30 meters but was badly hit by the depth charges. Significant damaged, she was forced to surface and abandoned. The 72 crew members, some of them hurt and bleeding, had to deal with heavy swells and sharks before being rescued by the destroyer *HMS Pakenham* and the corvette *HMS Jasmine* at about 9 in the morning. Twenty-four crew members perished.

The *Monge* was launched at La Seyne in June 1929 and commissioned 1932. She was deployed in the Mediterranean and in 1939 along Africa's west coast. She underwent a major careening in Toulon between October 1939 and April 1940 before proceeding to Bizerte (Tunisia). In October 1940 the *Monge* proceeded to Dakar and thence in December 1940 to Diego Suarez arriving in January 1941. Between March and September 1941, she sailed to Saigon (Indochina) returning to Diego Suarez in October. She quickly turned around to force the blockade and supply Djibouti returning to Diego Suarez on December 1, 1941.

The *Monge*, then captained by commander Delort, was in Reunion when Maerten issued the order to set sail for Diego Suarez. She cast off immediately, arriving from the east a few miles away from the entrance of the Bay of Diego Suarez during the night from 7 to 8 September 1942. Meanwhile, the full British squadron was sailing since the 7th in the morning from Courrier Bay bound for Diego Suarez

Figure 123. The Submarine *Monge*

Bay. On May 8 at 07:56 in the morning, at seven nautical miles east of the Anoronjia pass (the entrance to Diego Bay), the *Monge* attacked the aircraft carrier *HMS Indomitable*. But she only just avoided her torpedoes; one of which passed 50 yards in the front of the ship. The destroyers *HMS Active* and *Panther* immediately counterattacked and sank the *Monge* with all hands lost (5 officers and 64 sailors). Proof of the sinking included a piece of beam, a glove, a colonial helmet, some wooden debris and a large oil slick.

M-16B AND M-20B (JAPANESE MIDGET SUBMARINES)

Three weeks after the capture of Antsiranana, a Japanese submarine flotilla arrived at Diego Suarez. In a daring night raid they attacked the British ships in the bay. On March 27, 1942 the German naval staff had requested that the Imperial Japanese Navy launch operations against Allied convoys in the Indian Ocean to help relieve pressure on the Kriegsmarine. On April 8, the Japanese formally agreed to dispatch submarines to the east coast of Africa. The 1st Division of the 8th Submarine Flotilla was withdrawn from its base at Kwajalein in the Marshall Islands and arrived at Penang in northwestern Malaya in late April 1942. Commanded by Rear Admiral Ishizaki Noboru, the division was made up of Pearl Harbor veteran fleet submarines *I-10, I-16, I-18, I-20, I-27, I-28* and *I-30*. Three carried one midget type A Kai sub apiece and two carried one aircraft apiece. The submarines were supported by a pair of auxiliary cruisers/supply ships—*Aikoku Maru* and *Hokoku Maru*—armed with guns and torpedoes.

On the way to Diego Suarez, *I-30* was first to depart Penang on April 20, 1942 sending her scout plane over Aden's harbor on May 7 and then working southward with further reconnaissance at Djibouti, Mombassa, Dar-es-Salaam and Zanzibar. The main body took a more southerly course toward Durban where it undertook reconnaissance. As many as 40 Allied cargo ships lay in Durban roadstead but the undetected Japanese submarines, seeking warships, were not yet ready to show their hand and no attacks were launched. Instead, the subs concentrated off Diego Suarez. En route, all three "mother" submarines encountered heavy seas and were repeatedly swamped while charging batteries. *I-20* received medium damage which was quickly repaired. *I-18's* port diesel was flooded and four cylinders seized. As a result she and her midget submarine failed to reach Madagascar in time.

Around 10:30 pm on May 29, 1942 the *I-10's* floatplane reconnoitered the harbor at Diego Suarez. The plane sighted the admiral ship *Ramillies* at anchor in the bay, with two destroyers, two corvettes, a troopship, the hospital ship *Atlantis*, the tanker *British Loyalty*, a merchant and an ammunition ship. Although the scout plane was spotted—and *Ramillies* left her berth—the Japanese flight was believed to be by a French airplane from an airfield down-island. The floatplane returned to *I-10* unharmed.

Later that night, Captain Ishizaki ordered a midget submarine attack for the next day, *I-16, I-18,* and *I-20* were ordered to launch their midget subs. A little after 5 pm on May 30, at about 10 miles east of Diego Suarez, both *I-16* and *I-20* launched their midget sub-

Figure 124. The *I-16* submarine (Reproduced with the kind permission of B. Hackett and S. Kingsepp, From "Sensuikan," op. cit.)

marines (*M-16b* and *M-20b*) and penetrated the harbor. The craft from *I-18* was not successfully launched due to engine trouble. The *M-16b* was crewed by ensign Katsusuke Iwase as Captain and petty officer Takazo Takata as the navigator. The *M-20b* was crewed by Lt. Saburo Akieda as Captain and petty officer Takemoto as the navigator.

It was now nightfall on May 30 with clear sky, bright and full moon. At 8:25 pm, the British admiral ship *Ramillies* was hit by a torpedo of the midget submarine *M-20b* and severely damaged.[225] The torpedo holed the bulge and bottom plating opening a 30-foot by 30-foot hole in the port bulge forward of the "A" turret. The battleship's electrical system suffered damage and power was lost all over the ship. At 9:20 pm, while the corvettes *Genista* and *Thyme* dropped depth charges nearby, the *M-20b* fired her other torpedo and sank the 6,993-ton *British Loyalty* tanker in about 65 feet of water.[226] The British were caught totally off-guard. Prior to discovering the true agents of the attack Allied planes flew a retaliatory raid against the airfield at Antananarivo and claimed to have destroyed three French aircraft.

The midget submarine *M-16b* went missing in action but her wreck was never found. Some reports indicate that she was unsuccessful in getting into the bay. This is not supported by eyewitness reports from crewmembers on both the *Ramillies* and the *British Loyalty*. Both reported seeing the conning towers of two small submarines in the harbor around the time of the attack. On June 2, 1942 the body of an unidentified Imperial Japanese Navy sailor, presumably *M-16b's* Iwase or Takata, was found on a beach off Diego Suarez.

The *M-20b* escaped the bay unharmed and headed northnorthwest. After the midget's battery became depleted, *M-20b* was grounded on a reef at Nosy Antali Keli. Lt. Akieda attempted to scuttle his craft but the charge failed to explode. Both sailors reached shore and contacted local fishermen who arranged transport to the mainland. Akieda and Takemoto then headed for the recovery area near Amber Cape where the *I-20* was scheduled to wait for two days. In the morning of June 1 the two sailors visited Anijabe village to buy some food. While most villagers were friendly one contacted the British hoping to get a reward for his information. The next morning (June 2) at Amponkarana Bay, Lt. Akieda and Takemoto were intercepted by British Royal Marines after the Japanese had walked 48 miles. During an ensuing gunfight both Japanese sailors and one Marine were killed. After the other mother submarines departed the recovery area,

I-20 surfaced and unsuccessfully tried to contact the midget submarines by firing flares and sending radio signals. Not receiving any response, the *I-20* left the recovery area at 6:00 pm.

The limited remains of the *M-20b* were later sighted by British air reconnaissance and found about two weeks after the attack sitting upright on a reef within an area of volatile surf. The wreck is now completely submerged. The propellers were salvaged and are on display at the midget submariners' gravesite.

With Madagascar's main naval base now firmly in British hands it was thought that the French Governor General, Armand Annet, would surrender the whole island. However he refused and Britain was forced to mount further military operations. The second assault took place at Mahajanga on September 10, 1942 and was successful. Land forces started their march on Antananarivo. The assault forces returned to Diego Suarez and then sailed to Tamatave arriving on September 18. The landings were unopposed and the troops marched to Antananarivo which fell on September 23. However, Annet and his troops had retreated to the south. When a South African force landed at Tulear, Annet realized that he was trapped and surrendered on November 6, 1942 six months after the attack on Diego Suarez. The island was then handed to General de Gaulle's Free French movement. But the naval base at Diego Suarez remained under British control until 1944.

REFERENCES

[219] Sources: Grehan, J., "The Forgotten Invasion: The Story of Britain's First Large-Scale Combined Operation, the Invasion of Madagascar 1942," Historic Military Press, London, 2005; Hackett, B. and Kingsepp, S., "Sensuikan!," at http://www.combinedfleet.com/Madagascar.htm; Daubigny, B., "Le Drame de Diégo-Suarez," at http://perso.orange.fr/bertand.daubigny/Hdiego.htm; Rohwer, J., "Axis Submarines Successes of World War Two: German, Italian and Japanese Submarines Successes of World War II, 1939 to 1945," Annapolis, MD: Naval Institute Press, 1983; La Niece, P., "Madagascar Operations in World War II," in Madagascar, The Bradt Travel Guide, 7th Edition, Bucks, 2002, p. 326; Notices Historiques, Marine Nationale, Archives Centrales, Paris.

[220] For details about the *Bougainville*, see "Les batiments ayant porté le nom de Victor Schoelcher," at http://www.netmarine.net/bat/ae/schoelcher/ancien.htm.

[221] For details on the *Beveziers,* see Maître Godin, "Historique du Premier *Beveziers*" Actes du XXVè congrès des Sociétés Historiques et

Archéologiques de Normandie (October 4-7, 1990), communications de Vergé-Franeeschi, M. and Zesberg, A., at http://sous-marins.chez-alice.fr/historique_du_sous_marins_a_doub.htm.

[222] For details, see Turner, L., "Derrick and Sunderland T9040," Royal Air Force Archive List No. A5900861, at http://www.bbc.co.uk/ww2peopleswar/stories/61/a5900861.shtml.

[223] The *d'Entrecasteaux* was refloated in 1943, joined Aden and then Bizerte (Tunisia) in 1944, and was put out of service in October 1948.

[224] For more details, see Pénette, J.P. and Pénette Lohau, C., "Le Livre d'or de l'Aviation Malgache," Antananarivo, 2005, p. 27.

[225] After emergency repairs, *Ramillies* emerged from Diego Suarez on June 9 and limped to Durban, where she was docked for temporary repairs from June to August 1942. Later, *Ramillies* returned to Plymouth, England where full repairs were completed from September 1942 to June 1943. In June 1944, *Ramillies* participated in the D-Day invasion.

[226] *British Loyalty* was later refloated and towed to Addu Atoll (the southernmost atoll of the Maldives) where, on March 9, 1944, she was torpedoed and damaged by German submarine *U-183*. On January 15, 1946 *British Loyalty* was scuttled and sank off Addu Atoll. For more details see Butler, M., "The British Loyalty," 2002, at http://www.gan.philliptsmall.me.uk/Articles/BritishLoyalty.htm.

13

Other Wrecks off Madagascar

There are numerous other ships from all over the world beyond the ones portrayed above that wrecked in Madagascar's often treacherous waters from the beginning of ages to today. This section summarizes some of these wreckages. They are placed here because of want of more complete information. The list is far from being exhaustive.

HIRONDELLE[227] (1743)

The *Hirondelle* was a small 90-ton French ship mounting 6 cannons that left Lorient on her maiden voyage on September 17, 1731 under the command of Captain Antoine-Paul de Castillon accompanied by the ship l'*Oiseau*. After putting in at Goree in Senegal, she sailed to the Indian Ocean and was used as a supply ship between Bourbon, Mauritius and Madagascar between 1732 and late 1735. She left Bourbon for the return voyage in March 1736 arriving in Lorient in July. After being careened, *Hirondelle* undertook a second voyage the next year bound for India where she was used for a few years for coastal navigation. On her return voyage she foundered between Banivoule and Foulpointe in early February 1743. Her Captain and crew of 15 were picked up in Foulpointe by the *Fulvy* on February 14, 1743.

GLORIEUX[228] (1755)

The *Glorieux,* a 528-ton French frigate mounting 14 cannons was launched in St. Malo on July 26, 1750 arriving in Lorient on August 27 of the same year. She left for her maiden voyage on November 21, 1750 bound for Mauritius under the command of Captain Jean-Baptiste d'Après de Mannevillette and 89 crew members arriving on October 10, 1751. On November 21, 1751 her crew transferred to the *Treize-Cantons,* a 480-ton frigate commissioned by the French East India Company. The *Glorieux* was then used as a coastal ship in the southwestern Indian Ocean and condemned at St. Marie Island on July 22, 1755.

GANGE[229] (1762)

The *Gange,* a 600-ton French vessel built in 1753, mounted with 20 cannons and having a crew of 130 left Lorient on her maiden voyage bound for Bengal on November 19, 1754 under the command of Captain Antoine Tripier de Barmont. She was then used for coastal navigation between the various French trading posts in India, and left Bengal for the return trip to France on March 20, 1757. She undertook a second voyage to India leaving Lorient on January 22, 1759 under the command of Captain Julien de la Villebague with 142 crew members on board. She again did some coastal navigation in India, and subsequently loaded off in Mauritius on August 31, 1760. She was then used for coastal navigation in the southwestern Indian Ocean.

Between May and September, 1672, the *Gange,* under the command of Captain Desjardin, did several roundtrips between Bourbon or Mauritius and Foulpointe, essentially to procure cattle. In early October 1762, when lying at anchor at Foulpointe the quartermaster descended with a lantern in the hold which was full of sacks of hay. He accidentally dropped the lantern and the hay in the hold immediately caught fire. In no time, the whole ship was set ablaze. The crew jumped overboard; some swam to the coast and others were taken on board another vessel anchored there. The *Gange* quickly sank and all the cargo was lost.

L'HEUREUX[230] (1769)

The 300-ton l'*Heureux,* armed with 12 cannons, had been captured in India, together with the *Désir,* by the Saint-Georges squadron on January 20, 1762 and was chartered by the French East India Company in 1766. She left Lorient bound for India on March 7, 1767, under the command of Captain Jean Le Fer de Beauvais, with a crew of 56. She remained in India and was used for coastal navigation until April 1768, when she was sent to Mauritius. In 1769, the *Heureux* was sent by Pierre Poivre, then Governor of Mauritius, to Madagascar to solve a number of legal and administrative problems. Taken in a storm, she foundered in Foulpointe. The crew escaped and dispersed inland.

AMPHITRITE[231] (1799)

In April 1786, Charles René Magon de Medine became the Captain of this late vessel of the French East India Company. He occupied the Island of Diego Garcia with her for a second time. Then he undertook a hydrographic survey of the Seychelles. In 1794, the *Amphitrite* was used as a privateer in the Indian Ocean under the command of Captain François Le Meme. In 1797 she took a new Lieutenant, Jean Mallet who was Captain of the privateer *Le Mutin* that had captured the *Joachim* with the survivors of the *Winterton* in 1793.[232] The *Amphitrite* foundered off the western coast of Madagascar in 1799.

CHANCE (1799)

The *Chance* was a British merchant vessel. In the midst of the French Revolutionary War (1793-1802), laden with rice, she was cut out of Balasore Roads on India's east coast by the French frigate *Forte.* On April 25, 1799 *HMS Jupiter, Tremendous* and *Adamant,* which had sailed from Table Bay, recaptured the *Chance* off Mauritius. The *Chance* continued to ply the southwestern Indian Ocean. She was hit by a violent cyclone on November 5, 1799 and was lost off Madagascar.

LA DÉSIRÉE[233] (1810)

Letters from the Cape of Good Hope state the wreckage in October 1810 of a French corvette *La Désirée,* mounting 14 guns and with a crew of 112 men on the coast of Madagascar. *La Désirée* sailed from Mauritius for Europe with valuable freight. Of the ship's crew, 34 survived, only to be taken into slavery by the natives.

SAINT-VINCENT DE PAUL, D'APRÈS, AND MEUNIER[234] (1840)

On February 25 and 26, 1840 the southwestern Indian Ocean was hit by a ferocious cyclone. Six sailing ships (and some smaller boats) dragged on their anchors in the Port of St. Denis (Réunion) and started to drift in the raging sea towards Madagascar. Three were lost at sea but the *Saint-Vincent de Paul, D'Après* and *Meunier* came crashing on Madagascar's eastern coast between Mananjary and Mahela on February 28, 1840.

L'AUGUSTINE[235] (1855)

In the middle of the reign of Queen Ranavalona I and her fierce campaign against all foreigners, the French ship l'*Augustine* wrecked on Madagascar's east coast around August 1855. Five members of the crew were taken prisoners by the Hovas, accused of trying to hire workers for Reunion (which was punishable by death at the time) and taken to Antananarivo. The enterprising Mr. Laborde intervened and they were released against a ransom of 3,500 francs.

ANNE-MARIE[236] (1882)

The *Anne-Marie,* a French three-mast sailboat, was lost at Tamatave at Tanio Point.

LA BOURDONNAIS, MARGARETH AND IRENE[237] (1893)

Another cyclone hit Madagascar's east coast on February 28, 1893. The French cruiser *La Bourdonnais* was lost on the Madame islet at St. Marie with numerous lives lost. In Tamatave two German schooners were thrown on the reef: the 150-ton *Margareth* north of Tanio point and the 200-ton *Irene* on the large reef facing Tanio. That same storm also claimed one towing ship and 22 barges.

CONWAY CASTLE[238] (1893)

The 2,966-ton cargo steamship *Conway Castle* was built in 1877 by Robert Napier & Sons at Glasgow powered by a triple-expansion engine with a service speed of 12 knots. She was delivered for the mail service in September 1877 and became an intermediate steamer in 1883. On May 10, 1893 on a voyage from London to Mauritius, with nine passengers and 400 tons of freight, she hit the Vaudreuil reef off Vatoumandry 50 miles south of Tamatave. On the following day she was abandoned and became a total loss.

JOSEPH A. ROPES[239] (1894)

The American bark *Joseph A. Ropes* plied the route between Madagascar and New York for more than a decade. In the early 1890s she was used to secretly supply gunpowder to the Hova government in anticipation of the French desire to take possession of Madagascar. She was lost in Tamatave in 1894 opposite the mosque.

LOTSEN[240] (1894)

The Norwegian bark *Lotsen,* loaded with rubber, hit rocks at the point of Fort Dauphin in 1894 and was lost. Four years later, the chains of the French navy cruiser *Lapérouse* scraped her submerged wreck and ruptured causing the vessel to run aground.

MYRTLE M.[241] (1897)

The 151-ton Canadian sailboat *Myrtle M.* was thrown on the reef at Farafangana on Madagascar's southeast coast on May 6, 1897.

DRAGUETTA[242] (1898)

The Italian three mast barge *Draguetta,* under the command of Captain Scheffino, with 13 crew members and a cargo of salt and absinthe was thrown on the coast at Mahela (Tintingue, on the east coast) on July 2, 1898. There were no victims but the ship and cargo were lost.

ALOUETTE[243] (1898)

The French schooner *Alouette,* owned by Mr. Artaud of Nantes, was sailing from Mananjary to Tamatave with a cargo of rice, wine and hides. On October 16, 1898 she hit Hastie Point when entering the harbor of Tamatave and foundered. There were no casualties and part of the cargo was saved in particular the 28 barrels of excellent wine!

ESPÉRANCE[244] (1899)

The 107-ton French schooner *Espérance,* commanded by Captain Paul Félix, was sailing from Fort Dauphin bound for Tamatave with nine crew members and four passengers. During the night of 2 to 3 April 1899, beset by a massive downpour and very heavy seas, she hit a reef off Matitanana while hugging the coast. Efforts to dislodge her failed and she was abandoned. The crew and passengers were saved. The Captain was declared responsible for the wreckage.

FALCON HURST[245] (1900)

The 1,997-ton British coal bark *Falcon Hurst,* leased to the Messageries Maritimes, was sailing from Cardiff, U.K. to Diego Suarez with 26 crew members and a cargo of coal. She missed the entrance of Diego Bay in heavy seas; hit the rocks of Nosy Vory, one of the Barracouta Islands in the northeast, on June 22, 1900. Rocked by violent monsoon winds the ship split into two and sank. One sailor drowned.

NINA AND ACTIO[246] (1900)

The east coast of Madagascar endured very heavy weather during the latter part of July 1900, a kind of violent tidal wave. On July 21, the coastal sailboat *Nina,* which belonged to the trading house Tronchet in Tamatave, took refuge in the bay of Fort Dauphin. The violence of the storm soon broke her chains and threw her on the coast with her load of 88 tons of rice.

On July 25, the 520-ton Norwegian bark *Actio,* Captain Olsen, with 11 crew members and a cargo of salt and aniseed-flavored spirits, left her berth in the harbor of Mananjary (southeast coast) around 5 pm. Hit by the storm, she ran aground in the middle of a narrow channel 2 miles north of the harbor. The violent gale then thrust the ship towards shore, the violence of the sea cut her in two and she ended up beached a few hundred feet from the shoreline on a reef running parallel to the coast.

The provincial administrator immediately arrived in the disaster area with a rescue team. They managed, with difficulty, to run a cable between the wreck and the coast enabling comings and goings and in this manner to save part of the crew. The cable ruptured twice interrupting the salvage. Nonetheless, with energy and perseverance they managed in the dark to save the Captain and the first officer who were still desperately hanging unto the wreck. Two crew members died in the accident and all of the cargo was washed away.

ROGER[247] (1901)

In February 1901, the 79-ton brick-schooner *Roger* owned by the Comptoir Nord-Est de Madagascar sailed from Diego Suarez to Antalaha (northeast of Madagascar) with a varied cargo including 18 barrels of rum from Reunion. She foundered in the harbor of Antalaha due to a storm on February 19, 1901. The crew and cargo were saved but the ship was lost.

HERMANN[248] (1901)

The 7,444-ton Norwegian bark *Hermann,* leased by the Compagnie Lyonnaise de Madagascar, sailed in mid-1901 from Marseille bound for Andevoranto and Vatomandry on Madagascar's east coast with a cargo of limestone destined for the building of the railway. On June 20, 1901 in very heavy seas, she hit a reef off the harbor of Mahanoro (south of Tamatave) and immediately started taking in much water. The Captain, who was unfamiliar with the coast, decided to beach the ship which broke up. The crew and part of the cargo were saved but the *Hermann* was abandoned.

PRINCESSE[249] (1901)

The schooner *Princesse,* belonging to the Compagnie Lyonnaise de Madagascar, ran aground on the Faux Cap in the south of Madagascar in early November 1901.

ALTAÏ[250] (1902)

The Russian bark *Altaï*, transporting 1,500 tons of briquettes, beached on January 25, 1902 at the northern point of the Hastie reef south of Tamatave after hitting the madrepores when entering the pass. She was dislodged the next day by the *Nansen* which beached her again on Tanio beach in order to offload her cargo and eventually have her repaired. However, bad weather discombobulated these plans. Just two weeks later on February 11-12 1902, a massive cyclone passed through Tamatave. The wreck of the *Altaï* was entirely broken up and carried away by the furious waves. It is said that the black sand in front of Tamatave's hospital originates from its cargo of briquettes.

NANSEN[251] (1902)

The small French steamship *Nansen*, Captain Nadeau was bound for Brickaville on Madagascar's east coast with a load of railroad tracks. In late September 1902 her engine suddenly broke down while crossing the race at the mouth of the Iaroka River. With her propeller out of service, the sea pushed the *Nansen* towards shore and ran her aground north of the mouth of the river. When trying to dislodge her, heavy weather pushed her keel further in the sand. She was so deeply entrenched that the insurance company decided to halt the rescue operation.

ROMFORD[252] (1902)

The 3,035-ton English steamship *Romford* ran aground north of Vohemar on October 29, 1902 and was lost.

NORMAND MACLEOD (AND OTHERS)[253] (1902)

A massive cyclone hit Madagascar from December 9 to 12, 1902 traveling through about three quarters of the island. The hurricane first stroke the northeast coast then moved overland bending southward. It was felt as far as the Betsileo region on its way down before returning to the Indian Ocean. The end result of these four tragic days was a string of calamities on land as well as on the sea. A violent tidal

Figure 125. The *Nansen* wrecked on the beach at Andovoranto (Source: Revue de Madagascar, December 1902, p. 562)

wave, felt from Vohemar to Fort Dauphin, significantly lengthened the list of maritime disasters off Madagascar. Sail ships in particular were the main victims of the furious assault of the ocean against the coast. The schooner *Marcelle* ran aground north of Vohemar; the *Espérance* did likewise close to Antalaha, the *Grenouille* at Angontsy, the *Brésil* at Tatsianarana, the *Favorite* at Vinambe and the *Berthe* at Tamatave. At Rantabe, on the southern shore of the Bay of Antongil, a humungous wave tossed the coastal ship the *Furet* about thirty feet inland.

During the night of December 10-11 while trying to find shelter, the schooner *Renée* of the Compagnie Mareillaise de Madagascar was flung against the large reef guarding the entrance to the port of Tamatave. Pounded by the waves, it looked as if the ship, her Captain, the eight crew members and a passenger on board would never make it. The pilot, Joseph Serneaux, braving the raging storm and the enormous waves swung out and boarded the boat. With the help of seven courageous Malagasy sailors he waged an incredible fight for four hours and managed to bring back everybody safely to the wharf. During the same night, the British bark *Normand Macleod* ran aground in front of the residence in Tamatave. Fortunately, all these accidents listed so far did not cause any loss in life.

The same cannot be said of the *Louisiana* and the *Voltigeur* which vanished. The only trace was some bodies washed ashore by the sea at Cap Masoala. Also, the *Tanjona,* seeking shelter, foundered between Soanierana-Ivongo and Antsiraka (Pointe à Larrée) opposite Sainte-Marie Island. The Captain and a Malagasy were sucked under by the whirlpool. The first officer, the crew and the cook managed to resurface and to clinch to a wreck. All but two were then washed away by the furious waves. When getting close to shore one of the two survivors was devoured by a shark. One, named Toto, managed though totally exhausted, to get ashore in Anteviala in the Bay of Antongil and survive to tell the tale of this tragedy.

GERTRUDA GERARDA[254] (1903)

On February 3, 1903, a large Dutch bark the *Gertruda Gerarda,* built at the AA Krimpen wharf at Lek loaded with wood hit some sandbanks lying between Ankaramandihy and Fitanamalona some 12 miles from Manantenina on the southeast coast and ran aground. No survivors were found on the ship whose deck was absolutely barren. The crew consisted of 18 people including one woman. Either they all drowned or the absence of any of the ship's boats might indicate that they found shelter either onshore or on some passing ship.

BIRMAH[255] (1903)

Another cyclone hit Madagascar on March 21 and 22, 1903. On Saturday the 21st, a tidal wave was signaled in Tamatave. Six barges ran aground and the furious sea swallowed up a few small crafts. The situation in the harbor became so painful for the large ships that the steamer *Ville de Majunga* of the Compagnie Havraise Péninsulaire was forced to move to the open sea. The southeast coast was also badly thumped. In particular the Italian sailboat *Birmah* ran aground four miles north of Andavaka close to Andevoranto, on March 22, 1903.

MARIE-THÉRÈSE[256] (1904)

The 147-ton sailboat *Marie-Thérèse* hit the reef at the mouth of the Monombo River (Foulpointe) on December 30, 1904 and was lost.

BENGALIA[257] (1905)

In January 1905, the 7,661-ton German steamer *Bengalia,* built in 1898 and belonging to the Hamburg-America line was one of sixteen big German steamers which went to Madagascar with Welsh coal to supply the Russian squadron anchored in Nosy Be. Loaded with 9,000 tons of coal, she ran aground in heavy weather in the area close to the Bay of Sainte Luce. The crew was saved but with difficulty.

CAVALAIRE (IVOLINA)[258] (1926)

The *Cavalaire* was a 682-ton cargo ship built in 1908 by the Thornycroft yards in Woolston, United Kingdom for the Marina Mercante Argentina in Buenos Aires. She was sold in 1925 to Serra and Houreau, renamed *Ivolina* and used as a coaster in Madagascar. On March 6, 1926 she hit the reef south of the Pointe Blevec (the southernmost point of St. Marie Island) on and was lost. There were no casualties.

PYRITE[259] (1926)

The 476-ton French Navy patrol craft *Pyrite* was built in 1919 in the Loire yards at Nantes. In the afternoon of August 10, 1926 she hit a coral mound when entering the harbor of Tintingue (Manompana on the northeast coast). The ship was lost but there were no casualties.

BERIZIKY, SAINTE-ANNE, AMANDA, ALSACE AND TALISMAN[260] (1927)

The exceptionally violent cyclone hitting Tamatave on March 3, 1927 caused other casualties beyond the *Catinat* and the *Ville de Marseille II.*[261] On March 2 around four in the afternoon, the commander of the mixed sailboat *Beriziky* tried to get away from the impending storm. A commonplace incident prevented her to leave her berth: her anchor was gripped in the chains of the buoys. The operation had to be postponed to the next day. This delay became fatal for the small vessel from which only one passenger, clinging to the wreck, was saved.

The 357-ton steamer *Saint-Anne* ran aground just south of the *Catinat* and was lost. The sailboat *Elisabeth* ran aground at Tanio point. She was then pushed inland and was later found at the landmark of kilometer 6 on the road to Ivoloina 800 meters inland. She was later put back afloat. Other victims were the small Norwegian steamer *Amanda* and the towing ships *Alsace* and *Talisman*. The latter two were thrown ashore and were found more than two kilometers inland, one on the Melville road, the other at the end of the airport.

GUDRUN[262] (1927)

The 950-ton Norwegian cargo ship *Gudrun* was built at the Akers shipyard in Christiania (Oslo) for the Otto Thorensen Linie and named *May*. She changed owners several times and was renamed *Tresfond* in 1911, *Frierfjord* in 1915 and finally *Gudrun* in 1921. While anchored in the harbor of Tamatave a violent gale in the evening of July 21, 1927 ruptured her chains. She ran aground on the beach in front of the hotel Lagrave and was lost.

SITARA[263] (1929)

The English schooner *Sitara* hit the reef at Foulpointe on June 20, 1929 and was lost.

ALBERT-MORILLON[264] (1930)

The coaster *Albert-Morillon* of the Compagnie Côtière de Madagascar commanded by Captain Pellegrini, ran aground on Diego Island north of the entrance channel to the Bay of Diego Suarez, in 1930 and was lost.

MAFIA[265] (1930)

The 525-ton cargo coaster *Mafia* was built in 1910 at the Bertin Frères shipyard in Bezons. She was sold in 1913 and renamed *Rio S. Matheus*. She was then sold again in 1916 to F. Ward retaking her original name. On March 22, 1930 then under the command of Captain Legall, she encountered a forceful northeastern gale and ran aground at Fenerive on the east coast.

MADINA[266] (1930)

The English schooner *Madina* ran aground due to a windstorm on August 3, 1930 at Mandrisy (south of Cape Bellone, on the northeast coast).

VOLONTAIRE[267] (1942)

The schooner *Volontaire* was lost with all hands in a gale on November 10, 1942 off the coast of Maintirano (west coast).

FORT FRANKLIN[268] (1943)

The *Fort Franklin* was a 7,135-ton British steamer built by West Coast Shipbuilders Ltd., of Vancouver and delivered in December 1942 for US War Shipping Administration and lend-leased on bareboat charter to Dodd, Thomson & Co (the King Line) in London for the British Ministry of War Transport. On July 16, 1943 off eastern Madagascar at position 22° 36'S/51° 22'E while sailing independently on a voyage from Port Said and Aden to Lourenco Marques and Durban with 1,500 tons of salt as ballast, she was sunk by a torpedo from U-181 commanded by Wolfgang Lüth (part of the same fleet as the U-197, below). The Master Captain Thomas Witney Trott, 43 crew and 9 gunners landed at Mananjara. Two crew were lost.

U-BOOT 197 (SUBMARINE)[269] (1943)

The *U-197,* an IXD/2 type German submarine, was laid down in July 1941 at AG Weser in Bremen and commissioned in October 1942. As IXD/1 and IXD/2 type submarines carried large quantities of fuel it was decided to send them as far as Madagascar and probe the traffic there instead of the empty Cape area. A fleet of seven submarines arrived to the area in May 1943. The *U-197* under the command of KrvKpt. Robert Bartels joined them a month later. They took pre-arranged routes to survey the traffic 600 to 700 miles off the African coast. *U-177* carried out experiments with Bachstelze—a small helicopter with an observer towed by the U-boat. Thirty flights were made but not a single ship was sighted. Miserable results were achieved when patrolling off southeast Africa.

As a result, the U-boats moved eastwards to a position 600 miles south of Mauritius. Here the group was refueled and reprovisioned on June 23, 1943 from the surface tanker *Charlotte Schliemann* sent from Japan. Special security measures were undertaken. After replenishing, the *U-197* took a position south of the Mozambique Channel. Good results were achieved after the area relocation and a number of ships were sunk mainly off Madagascar. The *U-197* sunk three ships and damaged another. In mid August 1943 the last U-cruiser started the return passage.

On August 20, 1943 *U-197* was sunk south of Madagascar (in position 28.40S, 42.36E) by depth charges from two Catalinas of the 262 and 259 Squadron RAF before *U-196* and *U-181* arrived to the spot to give assistance. All hands were lost; a total of 67 officers and crew. The sinking of the *U-197* was probably due to information gained from the breaking of the German ENIGMA codes. Access to these codes was one of the most jealously guarded of Allied secrets and enabled Allied High Command to eavesdrop on German operational radio messages throughout most of the war.

REFERENCES

[227] Source: Logbook of Captain de Castillon, Archives Nationales, Paris, Marine, 4JJ86, No. 14.

[228] Sources : French Naval Archives (Service historique de la défense—Centre des archives de l'armement), at http://www.servicehistorique.sga.defense.gouv.fr/04histoire/dossierdushd/marine/indes/central.htm; Maurel, H., "Navires de la Compagnie des Indes allant aux/venant des Isles de l'Océan Indien," at http://perso.orange.fr/henri.maurel/cieind2.htm.

[229] Source : Letter from Mr. Barré, first mate aboard the *Gange,* dated October 10, 1762, Archives Nationales, Paris.

[230] Source: French Naval Archives (Service historique de la défense—Centre des archives de l'armement), at http://www.servicehistorique.sga.defense.gouv.fr. Not to be confused with the frigate *L'Heureuse,* chartered by the French East India Company, which sailed from Lorient bound for Mauritius on June 30, 1768 and foundered on Providence Island (part of the Farquhar archipelago to the north of Madagascar) on September 6, 1769.

[231] Sources: Pritchard, E.H., "The Struggle for Control of the China Trade during the Eighteenth Century," *The Pacific Historical Review,* Vol. 3, No. 3 (Sep., 1934), pp. 280-295.

[232] Hood, J., op. cit., p. 219; see above.

[233] Sources: Grocott, T., "Shipwrecks of the Revolutionary and Napoleonic Eras," London: Chatham, 1997, p. 295; The Times, January 15, 1811.

[234] Source: Lhuillier, M., « Naufrages, Echouements et Accidents de Navigation survenus à Madagascar, » Bulletin de l'Académie Malgache, T. XXIX (1949-1950), p.84.

[235] Pfeiffer, Ida, op. cit., pp. 42-43.

[236] Source: Lhuillier, M., op. cit., p. 84.

[237] Source : Lhuillier, M., op. cit., p. 84.

[238] Source: The Red Duster Website at http://www.red-duster.co.uk/UNION11.htm.

[239] Source : Lhuillier, M., op. cit., p. 84.

[240] Source : Lhuillier, M., op. cit., p. 84.

[241] Source : Lhuillier, M., op. cit., p. 84.

[242] Sources : Valette J. and Ratsimbazafy, A., op. cit., p. 1056; Lhuillier, M., op. cit., loc. cit.

[243] Source : Valette J. and Ratsimbazafy, A., op. cit., p. 1057.

[244] Source : Valette J. and Ratsimbazafy, A., op. cit., p. 1057.

[245] Sources : Valette J. and Ratsimbazafy, A., op. cit., p. 1059; Lhuillier, M., op. cit., p. 85.

[246] Source : Revue de Madagascar, September 1900, p. 606 and October 1900, p. 675; Valette J. and Ratsimbazafy, A., op. cit., p. 1059.

[247] Source: Valette J. and Ratsimbazafy, A., op. cit., p. 1060.

[248] Source: Revue de Madagascar, September 1901, pp. 700-701; Valette J. and Ratsimbazafy, A., op. cit., p. 1060.

[249] Source: Valette J. and Ratsimbazafy, A., op. cit., p. 1060.

[250] Source : Revue de Madagascar, April 1902, p. 335.

[251] Source : Revue de Madagascar, December 1902, pp. 560-562.

[252] Source: Lhuillier, M., op. cit., p. 85.

[253] Source: Revue de Madagascar, February 1903, p. 142, March 1903, p. 257, and November 1903, p. 444.

[254] Source: Revue de Madagascar, April 1903, pp. 348-349.

[255] Source: Revue de Madagascar, May 1903, pp. 449-452; Lhuillier, M., op. cit., p. 85.

[256] Source: Lhuillier, M., op. cit., loc. cit.

[257] Source: Revue de Madagascar, April 1905, p. 365; The *New York Times,* January 19, 1905.

[258] Source: Lhuillier, M., op. cit., loc. cit; Miramar ship index, op. cit.

[259] Source: Lhuillier, M., op. cit., p. 87; Miramar ship index, op. cit.
[260] Sources: « Histoire de Tamatave » at http://tamatave.ifrance.com/histoire/histoire.htm; Chauvin, J., op. cit., pp. 163-167; and Lhuillier, M., op. cit., loc. cit.
[261] See above, under Compagnie Havraise Péninsulaire.
[262] Source: Lhuillier, M., op. cit., loc. cit; Miramar ship index, op. cit.
[263] Source: Lhuillier, M., op. cit., loc. cit.
[264] Source: Lhuillier, M., op. cit., loc. cit.
[265] Source: Lhuillier, M., op. cit., loc. cit ; Miramar ship index, op. cit.
[266] Source: Lhuillier, M., op. cit., loc. cit.
[267] Source: Lhuillier, M., op. cit., loc. cit.
[268] Source : http://fortships.tripod.com/war_damagelosses.htm
[269] Sources: Norman, F., "Dark Sky, Deep Water, First Hand Experiences of the Anti-U-Boat War in WWII," 1997; Niestle, A., "German U-boat losses during World War II," United States Naval Institute, 1998; Kenneth, W., U-Boat operations of the Second World War," United States Naval Institute, 1998; http://uboat.net/boats/u197.htm.

Bibliography

ARCHIVES

Académie Malgache, Antananarivo

Archives Départementales de la Réunion, St. Denis, Réunion.

Archives des Missions Etrangères, Ministère des Affaires Etrangères, Paris.

Archives Nationales (royales et coloniales), Antananarivo

Biblioteca do Museu da Marinha, Lisbon

Bibliotecas Municipais de Lisboa, Central—Palacio Galveias, Lisbon.

Bibliotecas Municipais de Lisboa, Biblioteca Municipal de São Lázaro, Lisbon.

Bibliothèque Grandidier, Antananarivo.

British Library, London.

Centre des Archives d'Outre Mer, Aix en Provence.

Centre Historique des Archives Nationales, Paris.

Marine Nationale française, Archives Centrales, Paris

National Archives of the United Kingdom, Kew, London

Rijksarchief, Den Haag.

Service Historique de la Défense, Paris and Lorient.

Books and Articles

General Historical Works on Madagascar and the Indian Ocean

Allibert, C., "Flacourt, édition de 1661 annotée et présentée par Claude Allibert," Paris: Inalco-Karthala, 1995.

Axelson, E., "Southeast Africa: 1488-1530," London-New York-Toronto: Longmans and Green, 1940.

Bulletin du Comité de Madagascar, 1895 through 1898.

Chauvin, J., "Le Vieux Tamatave (1700-1936)," Tamatave: F. Sourd, 1945.

Copland, S., "A History of the Island of Madagascar," London: Burton & Smith, 1822.

Dahl, O.C., "Migration from Kalimantan to Madagascar," Norwegian University Press / The Institute for Comparative Research in Human Culture, 1991.

Descartes, Macé, "Histoire et géographie de Madagascar depuis la découverte de l'île en 1506 jusqu'au récit des derniers évènements de Tamatave," Paris: P. Bertrand, 1846.

Dunoyer de Segonzac, A., « Un Conquérant sous la Mer Henri-Germain Delauze », Paris: Buchet/Chastel, 1992.

Ferrand, G. (with revisions by Vérin), "Madagascar," in Bosworth, C.E., van Donzel, E., Lewis, B., and Pellat, Ch. (eds.), *The Encyclopedia of Islam,* Leiden: Brill, 1986, Vol. 5.

Ferrand, G., "Les Tribus Musulmanes du Sud-Est de Madagascar," Revue de Madagascar, June 1903, pp. 481-491.

Fichot, E., « Les Côtes de Madagascar », Revue Maritime, 1902.

Flacourt, E., "Histoire de la Grande Isle Madagascar," Paris: Alexandre Lesselin, 1658 (and second edition, Troyes: Nicolas Oudot, 1661).

Galli, H., "La Guerre à Madagascar-Histoire Anecdotique des Expéditions françaises de 1885 à 1895," Paris: Garnier Frères, s.d.

Godfrey, T., "Dive Maldives, A Guide to the Maldives Archipelago," Richmond (Victoria, Australia): Atoll, 1996.

Grandidier, A., "Histoire de la Géographie de Madagascar," Paris, 1883

Grandidier, A., Charles-Roux, Delhorbe, C., Froidevaux, H., and Grandidier, G., "Collection des ouvrages anciens concernant Madagascar," Paris, Comité de Madagascar, 1903.

Grant, Charles (viscount de Vaux), "History of Mauritius," London: G. and W. Nicol, 1801.

Le Chartier, H. and Pellegrin, G., « Madagascar depuis sa découverte jusqu'à nos jours », Paris : Jouvet, 1888.

Mack, J., "The land viewed from the sea," Azania, XLII, 2007.

Owen, Captain W.F.W., "Narrative of Voyages to explore the shores of Africa, Arabia and Madagascar; Performed in H.M. Ships Leven and Barracouta," (2 volumes), New York: Harper, 1833.

Pauliat, Louis, « Madagascar », Paris: Calmann Levy, 1884.

Pfeiffer, Ida, "Voyage à Madagascar," Paris: Hachette, 1881.

Revue de Madagascar, 1899 through 1959.

Richemont (Baron, de), P., "Documents sur la compagnie de Madagascar, précédés d'une notice historique," Paris : Challamel Ainé, 1867.

Van der Cruysse, D., « 'O navigants, o povres mathelotz' les échecs maritimes du XVIe siècle, » in Le Noble désir de courir le monde. Voyager en Asie au XVIIe siècle, Paris: Fayard, 2002.

Vérin, P., "The History of Civilisation in North Madagascar," Rotterdam: Balkema, 1986.

Vérin, P. and Heurtebize, "La trañovato de l'Anosy, première construction érigée par des Européens à Madagascar. Description et problèmes," Taloha 6, 1974.

Villars (Capitaine, de), "Madagascar 1638-1894. Etablissement des Français dans l'Ile," Paris: L. Fournier, 1912.

Portuguese East India Company

Albuquerque, F. de, "Commentarios do Grande Alfonso d'Albuquerque," Lisbon, 1557 (reedited in 1776).

Albuquerque, Luís de, "Livro das Armadas," Lisboa; Academia das Ciências de Lisboa, 1979.

Andrada, Francisco de, "Crónica del Rey Don João III," Lisbon, 1613 and Porto: Lello e Irmão, 1976.

Barros, João de, "Da Asia Portuguesa," Lisbon, 1778 edition.

Barros, João de, "Ásia. Dos feitos que os Portugueses fizeram no descobrimento e conquistas dos mares e terras do Oriente," Lisboa: Agência Geral das Colónias, 1945-46.

Bellec, F., Oliveira, R., and Michéa, H., "Naus, caravelas e galeões: na iconografia das descobertas A careira da Índia no século XVI: realto de uma vulgar viagem ao inferno A arquitectura naval e a expansão marítima portuguesa Princiais navios de século XVI," Lisboa: Quetzal, 1993.

Boxer, C.R., "An introduction to the História Trágico-Marítima," (reprint) from the Miscelânea de Estudos em honra do Professor Hernâni Cidade, Lisbon, 1957.

Boxer, C.R., "Further selections from the Tragic History of the Sea, 1559-1565," Cambridge University Press, 1967.

Canitrot, "Les portugais sur la côte orientale de Madagascar et en Anosy au XVIe siècle (1500-1613-1617)," Revue de l'Histoire des Colonies, 1921, pp. 203-238.

Castanheda, F. Lopes de, "Historia de los descobrimentos e Conquista da Índia pelos Portugueses," Lisbon, 1552 (reedited 1833 and Porto: Lello e Irmão, 1979).

Correia, Gaspar, "Historia do descobrimento e primeira conquistas da India," Lisbon, 1516.

Correia, Gaspar, "As Lendas da Índia," Lisbon, 1555 and Porto: Lello e Irmão, 1975.

Correia, Gaspar, "Crónicas de Don Manuel e de Don João III," Academia das Ciências, Lisboa, 1992.

Couto, Diogo do, "Décadas da Asia, IV-VII," Lisbon, 1602-16

Couto, Diogo do, "Décadas" Vol. II; Lisboa: Livraria Sá da Costa

Cruz, Maria Augusta Lima, "As viagens extraordinárias pela rota do cabo" in *Actas do VII Seminário Internacional de História Indo Portuguesa;* Angra do Heroísmo; [s.n.], 1998.

da Costa Quintella, Ignacio, "Annaes de Marinha Portuguesa," 1839.

Decary, Raymond, "Les voyages des Portugais à Madagascar au 16e siècle," Paris, 1949.

Duffy, J., "Shipwreck and Empire. Being an account of Portuguese maritime disasters in a century of decline," Harvard University Press, London: Geoffrey Cumberlege, 1955.

Falcão, Luís, "Livro de toda a fazenda"; Lisboa: Imprensa Nacional, 1859.

Faria y Sousa, Manoel de, "Asia portuguesa," Lisbon, 1666.

Fonseca, Quirino da, "Os Portugueses no mar: Memórias Históricas e arqueológicas das Naus de Portugal," 1º Volume, Lisboa: Associação dos Arqueólogos Portugueses, 1983.

Gomes de Brito, B., "História Trágico-Marítima. Em que se escrevem chronologicamente os Naufragios que tiverão as Naos de Portugal, depois que se poz em exercicio a Navegação da India," Lisbon, 1735-6.

Grandidier, Alfred, "Histoire de la Découverte de l'Ile de Madagascar par les Portugais," Paris: C. Lamy, 1902 (excerpt from the *Revue de Madagascar* of January 10, 1902).

Guinote, P., Furtoso, E. and Lopes, A., "O movimento da carreira da Índia nos séculos XVI—XVIII. Revisão e propostas." in *Maré Liberum;* Lisboa: Comissão Nacional para a Comemoração dos Descobrimentos Portugueses; nº 4, 1992.

Guinote, P., Frutuoso, E., and Lopes, A., "Naufragios e Outras Perdas da Carreira da India: Séculos XVI e XVII" Lisboa: Grupo de Trabalho do Ministério da Educação para as Comemorações dos Descobrimentos Portugueses, 1998.

Kammerer, A., "La Découverte de Madagascar par les Portugais et la cartographie de l'île: 1500-1667," Boletim da Sociedade de Geografia de Lisboa, Série 67, Nº 9-10, 1949.

Lafitau, P., "Histoire des découvertes et conquêtes des Portugais dans le Nouveau Monde," 1733.

Maffei, P., "Historiarum Indicarum," Lyon, 1637.

Osorio, J., "De rebus Emmanuelis," 1574, translated in 1804, "Da vida e feitas del Rei Dom Manoel," XII livros.

Passos, Carlos dos, "Navegação Portuguesa dos Séculos XVI e XVII" in *O Instituto,* volume 64, pg. 85- 88; Coimbra; Imprensa da Universidade, 1917.

Peres, Damião, "Viagens e naufrágios célebres dos séculos XVI, XVII, XVIII," Porto: Alberto de Oliveira Ltd, 1937.

Sousa, Frei Luís de, "Anais de D. João III," Livro 4º, Lisboa: Livraria Sá da Costa, 1938.

Stenuit, R., "Ces Mondes Secrets où j'ai plongé," Paris: Robert Laffont, 1988.

Stenuit, R., "La Vergine di Monte Carmo," *Mondo Sommerso*, 1991, 352: 54-63

Stenuit, R., "Nossa Senhora do Monte do Carmo," *Tauchen*, 1992, 7: 52-58.

Vasconcellos, Frazão de, "Armadas da Carreira da Índia de 1500 a 1590: relação de Duarte Gomes Solois," Lisboa [s.n.], 1938.

Xavier, Padre Manuel, "Compêndio universal de todos os vice-reis, governadores, capitães geraes, capitães mores, capitães de naus, galeões, urcas, caravellas, que partiram de Lisboa para a Índia Oriental e tornaram da Índia para Portugal, com nomes de todos, dias, mezes e horas que partirão deste reino," Nova Goa: Imprensa Nacional, 1917.

Xavier, Padre Manuel, "Relações da Carreira da Índia," Dirigido e comentado por Luís de Albuquerque, Lisboa: Alfa, 1990.

Dutch East India Company

Bontekoe, W.Y, "Het Journael of de gedenkwaerdige beschrijvinghe and de Oost-Indische Reyse van W.Y. Bontekoe van Hoorn, begrijpende veel wonderlijke en gevaerlijcke saecken hem daar in wedervaren," Amsterdam, 1630.

Bostoen, K., Daalder, R., Roeper, V., Verhoeven, G., and Wildeman, D., "Bontekoe. De schipper, het journaal, de scheepsjongens," Zutphen, 1996.

Bruijn, J.R., Gaastra, F.S., Schöffer, I. "Dutch-Asiatic Shipping in the 17th and 18th Centuries," The Hague, 1979, 1987.

"Dagh-Register gehouden int Casteel Batavia vant passerende daer ter plaetse als over geheel Nederlandts-India," Batavia, Anno 1624-1673.

De Constantin, « Recueil des voyages qui ont servi à l'établissement et aux progrès de Compagnie des Indes Orientales formée dans les Provinces Unies des Pays-Bas », 1725.

Leibbrandt, H.C.V., "Précis of the Archives of the Cape of Good Hope: Riebeeck's journal," Cape Town: W. A. Richards & Sons, 1897.

Leupe, "Verhandelingen en Berigten bettrekkelijk het zeewezen en de zeevaartkunde; verzameld en uitgegeven door Jacob Swart," ch. XXII, "De vestiging der Hollanders op Mauritius in 1638," Amsterdam: Hulst van Keulen, new series, 1854.

Meilink-Roelofsz, M. A. P. 1980, "The structures of trade in Asia in the sixteenth and seventeenth centuries. Niels Steensgaard's «Carracks, caravans and companies». The Asia trade revolutions. A critical appraisal" in *Mare Luso-Indicum. L'Océan Indien, les pays riverains et les relations internationales. XVIe-XVIIIe siècles;* Paris; Société d'Histoire de l'Orient; vol IV, 1962.

Miller, R., "De Oostindievaarders," Amsterdam: Time-Life Boeken, 1981

Mulder, W.Z., "Hollanders in Hirado, 1597-1641," Haarlem: Fibula-Van Dishoeck, 1985;

RGP-GS166, "Dutch-Asiatic Shipping in the 17th and 18th centuries, Volume II, Outward-bound voyages from the Netherlands to Asia and the Cape (1595-1794)," Den Haag: Martinus Nijhoff, 1979;

RGP-GS167, "Dutch-Asiatic Shipping in the 17th and 18th centuries, Volume III, Homeward-bound voyages from Asia and the Cape to the Netherlands (1597-1795)," Den Haag: Martinus Nijhoff, 1979;

s.n., "Het tweede Boeck, Joernael uit Dagh-register inhoudende een warachtig verhael ende historische vertellinghe van de reyse, maart 1598," ed. Bernard Langenes, Middelbugh, 1601.

Valentijn, F., "Oud en Nieuw oost-Indiën, deel IV/A," Franeker: Van Wijnen, 2003.

van Dam, Pieter, "Beschrijvinge van de Oostindische Compagnie," 's-Gravenhage: Martinus Nijhoff, 1927

Worden, N., Van Heyningen, E., Bickford-Smith, V., "Cape Town: The Making of a City: An Illustrated Social History," Cape Town: Verloren, 1998.

English East India Company

Anthony J., "Catalogue of East India Company Ships' Journals and Logs 1600-1834," British Library, 1999.

[Buchan, George], "Narrative of the loss of the Winterton East Indiaman, wrecked on the coast of Madagascar in 1792; and of the sufferings connected with that event. To which is subjoined, a short account of the natives of Madagascar, with suggestions as to their civilization," by a passenger in the ship, Edinburgh, 1820.

Dale, John, "A Narrative of the Loss of the Winterton," C. Withington for B. Farrington, c. 1796.

Desperthes, « Perte du Vaisseau de la Compagnie des Indes, Le Degrave, sur la Côte de Madagascar, en 1701, et aventures de Robert Drury, » Ed. Eyriès, 1821.

Drury, R., "Madagascar: or Robert Drury's Journal during fifteen years captivity on that island," London: W. Meadows, 1729; reprinted in 1807 by Stodart and Craggs, Hull.

Hackman, Rowan, "Ships of the East India Company," World Ship Society, 2001.

Hakluyt, R., "The principal Navigations, Voyages, Traffick, Discoveries of the English," London, 1599.

Hackluyt, R., "Collection Early Voyages," London, 1810.

Hakluyt Society, "The Voyages of Sir James Lancaster, Kt., to the East Indies, with Abstracts of Journals of Voyages to the East Indies, during the Seventeenth Century, preserved in the India Office. And the Voyage of Captain John Knight (1606), to seek the Northwest Passage," Edited by Clements R. Markham, C.B., F.R.S., 1877.

Hakluyt Society, "The Hawkins' Voyages during the Reigns of Henry VIII, Queen Elizabeth, and James I," Edited, with an Introduction, by Clements R. Markham, C.B., F.R.S.,1878.

Hood, Jean, "Marked for Misfortune: An Epic Tale of Human Endeavour and Survival in the Age of Sail," London: Conway Maritime Press, 2003.

Keay, J., "The Honorable Company—A History of the English East India Company," HarperCollins, London, 1991.

Kerr, R., "A General History and Collection of Voyages and Travels Arranged in Systematic Order: Forming a Complete History of the Origin and Progress of Navigation, Discovery, and Commerce, by Sea and Land, from the Earliest Ages to the Present Time," 1813.

Owen, Captain W.F.W., "Narrative of Voyages to Explore the Shores of Africa, Arabia, and Madagascar; Performed in H.M. Ships *Leven* and *Barracouta*," New York: J & J Harper, 1833.

Pearson, M.P. and Godden, K., "In Search of the Red Slave: shipwreck and captivity in Madagascar," Sutton, 2002.

Secord, A.W., "Robert Drury's Journal and Other Studies," Urbana: Illinois University Press, 1962.

Sutton, Jean, "Lords of the East: The East India Company and its Ships 1600-1874," London: Conway Maritime Press, 2000.

French East India Company

Benyowsky, M.A., "Memoirs and Travels of Mauritius Augustus, Count of Benyowsky, Magnate of the Kingdoms of Hungary and Poland, one of the Chiefs of the Confederation of Poland, Consisting of his military operations in Poland, his exile into Kamchatka, his escape and voyage from that peninsula through the northern Pacific Ocean, touching at Japan and Formosa, to Canton in China, with an account of the French settlement he was appointed to form upon the island of Madagascar," Translated from the original manuscript by William Nicholson, London: G.G.J. and J. Robinson, 1790.

Benyowsky, M.A., "Voyages et Mémoires de Maurice Auguste, comte de Benyowsky, magnat des Royaumes de Hongrie et de Pologne, contenant ses opérations militaires en Pologne, son exil au Kamtchaka, son évasion et son voyage à travers l'Océan Pacifique, au Japon, à Formose, à Canton en Chine, et les détails de l'établissement qu'il fut chargé par le ministère français d'organiser à Madagascar," Paris : F. Buisson, 1791.

Bouchard, Paul, « "Malheurs et chimères: de la Rochelle à Madagascar au XVIIe siècle," Le Mois Littéraire et Pittoresque, Paris, 1906, No. 89, pp. 522-536.

Cauche, F., (collected by Morisot), "Relation véritables et curieuses de l'Ile de Madagascar; Relation du Voyage que François Cauche de Rouen a fait en l'Ile de Madagscar," Paris: Augustin Courbé, 1651.

Cotain, "Shipwreck of the *Indienne*," in Tales of Shipwrecks and Peril, London: Burns and Lambert, 1858.

Desperthes, J.L., "Histoire des Naufrages ou recueil des relations les plus intéressantes des naufrages arrivés depuis le XVème siècle jusqu'à nos jours. Nouvelle éd., refondue, corrigée et augmentée, by J.B.B. Eyriés," Paris : Ledoux et Touré, 1815.

Dufresne de Francheville, "Histoire de la Compagnie des Indes," 1740.

Froidevaux, Henri, "Les Préludes de l'Intervention française à Madagascar au XVIIe siècle, Navigateurs, géographes et commerçants français de 1504 à 1640," Revue des Questions Historiques, 1909, pp. 1-44.

Gaubert (Lieutenant), "François Cauche," Revue de Madagascar, 1903, April, pp. 289-305, and May, pp. 385-403.

Glazemaker, J.H. (trad.), "De Rampspoedige Scheepvaart der Franschen naar Oostindien onder 't beleit van Generaal Augustyn van Beaulieu, met drie Schepen, uit Normandyen," Amsterdam: Jan Rieuwertersz en Pieter Arentsz, 1669.

Gravier, Gabriel, "La compagnie orientale à Madagascar (1642-1672)," s.l.

Hutchinson, E.W., "Aventuriers au Siam au XVIIe siècle," [translated by H. Berland], Bulletin de la Société des études indochinoises, Vol. XXII, Saigon, 1947.

Kaeppelin, "Les Escales françaises sur la Route de l'Inde, 1638-1731"

Lanier, L., "Etude historique sur les relations de la France et du Royaume de Siam de 1662 à 1703," Versailles : E. Aubert, 1883.

Launay, A., "Histoire de la Mission de Siam 1662-1811, Documents Historiques," Paris, 1920.

Le Bris, Michel, "D'or, de rêves et de sang, L'épopée de la flibuste (1494-1588) ," Paris: Hachette, 2001

Lougnon, A., "Correspondence du Conseil Supérieur de Bourbon et de la Compagnie des Indes, 22 janvier 1724—30 décembre 1731," St. Denis (Réunion), 1934.

Lougnon, A., "Correspondence du Conseil Supérieur de Bourbon et de la Compagnie des Indes, 23/1/1736 au 9/5/1741," Paris, 1935.

Lougnon, A., "Vaisseaux et traites aux îles depuis 1741 jusqu'à 1746," Recueil Trimestriel de Documents et Travaux Inédits pour servir à l'Histoire des Mascareignes Françaises, Tome No. 1, Tananarive, Avril-Juin 1940.

Magry, P., "Journal d'une Navigation des Dieppois dans les Mers Orientales sous François 1er (1529-1530)," Bulletin de la Société Normande de Géographie, May-June and July 1883, pp. 168-248.

Martin, Fr., "Extraits des Mémoires sur l'Etablissement des Colonies Françoises aux Indes Orientales, 1665-1668," Manuscript, Bibiliothèque Grandidier, Antananarivo, No. 2987.

Martin de Vitré, F., "Premier Voyage fait aux Indes Orientales par les Français, en 1602,"

Pauliat, Louis, « Louis XIV et la compagnie des Indes Orientales de 1664, d'après des documents inédits tirés des archives coloniales du ministère de la marine et des colonies ». Paris : Calmann Levy, 1886.

Roncière, de la C., « Histoire de la Marine française, » 1909.

Ruelle, « Voyage à Madagascar et aux Indes orientales, » s.d., manuscript, Bibliothèque Grandidier, Antananarivo, No. 2723.

s.n., "Quelques documents concernant la perte du négrier *Vautour* à Madagascar, en 1725," Recueil Trimestriel de Documents et Travaux Inédits pour servir à l'histoire des Mascareignes françaises, Janvier-Mars 1937, No. 4, pp. 347-372.

Souchu de Rennefort, « Histoire des Indes Orientales, » Leide : Frederik Harring, 1688.

Vacher, P., « Contribution à l'histoire de l'établissement français à Madagascar par le Baron de Benyowszky (1772-1776), D'après de nouvelles sources manuscrites », Collection 'Clio en Afrique,' N° 19, été 2006, Editions du Centre d'Etude des Mondes Africains, MMSH, Aix en Provence.

Pirates

Bibique, « Sur la piste des frères de la côte, » Paris: Orphie, 1988.

Bolster, W. J., "Black Jacks: African American Seamen in the Age of Sail," Harvard University Press, 1997.

Brockway, T., "The pirates in Madagascar," Antananarivo Annual, 1886.

Bucquoy, de J., "Zestien Jaarige Reize naar de Indiën gedaan door Jacob de Bucquoy, vol Aanmerkelyke Ontmoetingen," Haarlem: Jan Bosch, 1758 (second ed.).

Burg, B. R., "Sodomy and the Pirate's Tradition: English Sea Rovers in the Seventeenth Century Caribbean," New York, 1984.

Clifford, B., "Return to Treasure Island and the Search for Captain Kidd," New York: Harper Collins, 2003.

Cordingly, D., "Life among the Pirates: The Romance and the Reality," London: Little Brown & Co., 1995.

Dampier, W., "A New Voyage Round the World," London, 1697.

Deschamps, H., "Les Pirates à Madagascar aux XVIIème et XVIIIème Siècles," Paris: Berger-Levrault, 1949 (and second edition, 1972).

Downing, Cl., "A Compendious History of the Indian Wars," 1737.

Guët, I., "Les Origines de l'île Bourbon et de la colonisation française à Madagascar," 1886.

Johnson, Ch. (Captain), "General History of the Robberies and Murders of the most notorious Pyrates," London: J.M. Dent & Sons, 1724.

Lizé, P., "Piracy in the Indian Ocean, Mauritius and the Pirate Ship *Speaker*," in Skowronek, R.K. and Ewen, C.R. (Editors), "X Marks the Sport, The Archaeology of Piracy," University press of Florida, Gainsville, 2006.

Lougnon, A., « Sous le signe de la Tortue, voyages anciens à l'île Bourbon (1611-1725) », Paris : Azalées Editions, 1992.

Merveille, de la (Captain), « Mémoire manuscrit » addressed to Minister de Pontchartrain, 1712, French Archives Coloniales, Correspondance générale, Madagascar.

Molet-Sauvaget, A., "Un Européen, roi 'légitime' de Fort Daupin au XVIIIe siècle: le pirate Abraham Samuel," Etudes Océan Indien, 1997, 23/24: 211-221.

Ovington, J., "A voyage to Suratt in 1689."

Pouillaude, D., "Le Grand Livre des Aventuriers des Mers, Pirates, flibustiers, boucaniers, corsaires . . .," Orphie, 2005.

Sakolsky and Koehnline (eds.), "Gone to Croatan: The Origins of North American Dropout Culture," New York/Edinburgh: Autonomedia/AK Press, 1993.

Snelgrave, "A new Account of some parts of Guinea and the Slave-trade," 1734.

Viala, G., « La Buse, un pirate dans l'Océan Indien, » Ed. Du paille-en-queue noir, 1997.

Zacks, R., "The Pirate Hunter, The True Story of Captain Kidd," New York, 2002.

British Royal Navy

Hepper, D. J., "British Warship Losses in the Age of Sail 1650-1859," Sussex: Jean Boudriot, 1994.

Huntress, K., "Checklist of Narratives of Shipwrecks & Disasters at Sea to 1860," Iowa State University Press, 1979.

Gosset, W. P., "Lost Ships of the Royal Navy, 1793-1900," London: Mansell, 1986.

Grocott, T., "Shipwrecks of the Revolutionary and Napoleonic Eras," London: Chatham, 1997 and London: Caxton, 2002.

James, W., "The Naval History of Great Britain from the Declaration of War by France in February 1793 to the Ascension of George the IV in January 1820," London: Harding, Lepard & Co., 1826.

Norie, J. W. (compiler), "The Naval Gazetteer, Biographer and Chronologist, Containing a history of great Wars from their commencement in 1793 to their conclusion in 1801; and from recommencement in 1803 to their final conclusion in 1815," London: J.W. Norie & Co., 1827.

French Navy

s.n., "Voyage du Général Gallieni (Cinq mois autour de Madagascar)," Le Tour du Monde, 1899-1900.

Zurcher, Frédéric, and Margollé, Elie, « Récits des Naufrages célèbres avant 1900 », 5th ed., Paris: Hachette, 1888, Ch. 16, "Naufrage du brick le Colibri en vue des îles Radama (Madagascar) (1843)," pp.169-179.

Sailing Vessels and Steamers of the 19th Century

Cotain (Rev.) in "Tales of Shipwrecks and Peril," London: Burns and Lambert, 1858.

Jacobs, T. J., "Scenes, Incidents, and Adventures in the Pacific Ocean, or the Islands of the Australian Seas, during the cruise of the clipper Margaret Oakley, under Capt. Benjamin Morrell," New York: Harper and Brothers, 1844.

Lisagor, P. and Higgins, M., "Overtime in Heaven: Adventures in the Foreign Service," Garden City: Doubleday, 1964, Ch. 3, pp. 45-62.

s.n., "Le cyclone de Tamatave," La Nature, 1888, pp. 317-318.

Valette, J., « Le naufrage de l'Elisa à Tamatave, le 26 novembre 1827 », Bulletin de Madagascar, No. 254-255, July-August 1967, pp. 533-536.

Russian Fleet 1904-1905

Machikin, E., "Report on the Stay of Second Pacific Squadron in the Bay of Nosy-Be in Madagascar in 1904-1905," Research Historical Group of the Naval Fleet, Moscow, 2004.

Piouffre, G., "La guerre Russo-Japonaise sur mer," Nantes: Marine éditions, 1999.

Ranurusehenu, H., "The Stay of the Russian Fleet in Madagascar in 1904-1905," Études Océan Indien, 18, 1994.

Compagnie des Messageries Maritimes

Bois, P., « Le grand siècle des Messageries Maritimes », Chambre de Commerce et d'Industrie de Marseille-Provence, 2ème édition, 1992.

Campiniano-Cantemir, J., « La Fin du Salazie, par un Naufragé, » Paris : Chamolle, 1913.

Lanfant (Commandant), « Historique de la Flotte des Messageries maritimes-1851-1975, » Association des Anciens des Etats-majors des Messageries Maritimes, second ed., 2001.

Saibène, M., Brouard, J.Y., and Mercier, G., "La Marine Marchande Française 1940/1942, » Vol. II.

Compagnie Havraise Péninsulaire

Limonier, C., « Les 110 Ans de la Havraise Péninsulaire, Histoire de la Flotte », Marseille: P. Tacussel, 1992.

WWII—Diego Suarez

Grehan, J., "The Forgotten Invasion: The Story of Britain's First Large-Scale Combined Operation, the Invasion of Madagascar 1942," Historic Military Press, London, 2005.

La Niece, P., "Madagascar Operations in World War II," in Madagascar, The Bradt Travel Guide, 7th Edition, Bucks, 2002, p. 326.

Pénette, J.P. et Pénette Lohau, C., "Le Livre d'or de l'Aviation Malgache," Antananarivo, 2005.

Rohwer, J., "Axis Submarines Successes of World War Two: German, Italian and Japanese Submarines Successes of World War II, 1939 to 1945," Annapolis, MD: Naval Institute Press, 1983.

Other Wreckages

Jacomy-Régnier, « Colonie auvergnate à Madagascar », Revue d'Auvergne, t. 1, March 1840-1841, pp. 166-174.

Jacomy-Régnier, « A la recherche d'une colonie perdue », Marine de France, No. 6, January 25, 1895.

Kenneth, W., "U-Boat operations of the Second World War," United States Naval Institute, 1998.

Lhuillier, M., "Naufrages, Echouements et Accidents de Navigation survenus à Madagascar," Bulletin de l'Académie Malgache, T. XXIX (1949-1950), pp. 84-88.

Meyniard, C., "A la recherche d'une colonie perdue-Les Auvergnats à Madagascar," Bulletin de la Société de Géographie de l'Est, 1905, pp. 52-55.

Niestle, A., "German U-boat losses during World War II," United States Naval Institute, 1998.

Norman, F., "Dark Sky, Deep Water, First Hand Experiences of the Anti-U-Boat War in WWII," 1997.

Valette J. and Ratsimbazafy, A., « Les naufrages et échouements survenus à Madagascar de 1898 à 1901, » Bulletin de Madagascar, No. 235, December 1965, pp. 1056-1060.

van Spaan, G., "De Gelukzoeker over Zee of D'Afrikaansche Weg-Wijzer," Rotterdam: Pieter vander Slaat, 1694.

WEBSITES

General Historical Works on Madagascar

Beck, S., Africa and Slavery 1500-1800, in "Middle East & Africa to 1875." at http://san.beck.org/1-13-Africa1500-1800.html#11

Gilbert, William, *Renaissance and Reformation*. Lawrence, KS: Carrie, 1998, Chapter 10, "Exploration and the discovery beginnings of the expansion of Europe," formatted and installed as an e–book by Lynn H. Nelson at the University of

Kansas, at http://vlib.iue.it/carrie/texts/carrie_books/gilbert/10.html.

Wikipedia, History of Madagascar, http://en.wikipedia.org/wiki/History of_Madagascar.

Portuguese East India Company

Castro, F., 2006, India Route Ships Project: Introduction, World Wide Web, URL, http://nautarch.tamu.edu/shiplab/, Nautical Archaeology Program, Texas A&M University.

Furtoso, E., "India Route Project: *Relação de Capitaens Mores e Naos que Vierão do Reyno a este Estado da India des do seu Descobrimento,*" WWW, URL, http://nautarch.tamu.edu/shiplab/, Nautical Archaeology Program, Texas A&M University, 2003

Guinote, Paulo J.A., "Ascensão e Declinio da Carreira da India (Seculos XV-XIII)," at http://nautarch.tamu.edu/SHIPLAB/01guifrulopes/Pguinote-nauparis.htm.

Guinote, Paulo J. A., "Índia Route Project: Armadas que partiram para a Índia (1509—1640)," Biblioteca Nacional de Lisboa; Reservados; Caixa 26, nº153, 2003 WWW, URL, http://nautarch.tamu.edu/shiplab/

Maldonado, M. H., "Introdução a Relação Das Náos e Armadas da India Com os successos dellas que se puderam saber, Para Noticia e instrucção dos curiozos, e amantes Da Historia da India," leitura e anotações de Maria Hermínia Maldonado, Coimbra: Biblioteca da Universidade, 1985, at www, url, carreiradaindia.net/index.php?paged=2&s=moor)

Pissarra, J.V.A., "Navegações Portuguesas, Sousa, Pêro Lopes de," 2002, at WWW, URL, http://www.instituto-camoes.pt/cvc/navegaport/g61.html.

Dutch East India Company

Major, R.H., "Early Voyages to Terra Australis, now called Australia," a Gutenberg of Australia eBook No. 0600361.txt produced by: Col Choat and Bob Forsyth, http://gutenberg.net.au, April 2006.

Sabrizain, "VOC: The first multinational" in Sejarah Melayu, A history of the Malay Peninsula, at http://www.sabrizain.demon.co.uk/malaya/dutch.htm.

The VOC site, http://www.vocsite.nl/

English East India Company

British East India Company, Wikipedia at http://en.wikipedia.org/wiki/British_East_India_Company.

Brooke: History of St. Helena, at http://www.bweaver.nom.sh/brooke/brooke_ch2.html.

Edwards, R. "Robert Pitcairn," 2004, at http://www.findagrave.com/cgi-bin/fg.cgi?page=gr&GRid=9288829

Lettens, J., "HMS Aurora" at http://www.wrecksite.eu/wreck.aspx?16210

Pearson, M.P., "Shipwreck into Slavery," British Archaeology Journal, Issue 67, October 2002 at http://www.britarch.ac.uk/ba/ba67/feat2.shtml.

Phillips, M., "Ships of the Old Navy" at http://www.ageofnelson.org/MichaelPhillips/info.php?ref=0255

Sheffield University: "Bibliography for Robert Drury, his life and journal" at http://www.shef.ac.uk/archaeology/research/madagascar/robert_drury

French East India Company

Eschapasse, B., « Les Français chassés du Paradis, » Historia, at http://www.historia.presse.fr/data/mag/722/72202601.html.

French East India Company, Wikipedia at http://fr.wikipedia.org/wiki/Compagnie_fran%C3%A7aise_des_Indes_orientales.

French Naval Archives (Service historique de la défense—Centre des archives de l'armement), at http://www.servicehistorique.sga.defense.gouv.fr/04histoire/dossierdushd/marine/indes/central.htm;

Hallstrom, S., Shipwreck Explorer-Soleil d'Orient at http://www.shipwreckexplorer.com/hallstrom/soleil/soleil.htm;

Histoire de la Compagnie des Indes at http://enguerrand.gourong.free.fr/oceanindien/p01oceanindien.htm.

Le Lan, J-Y., "Le commerce français avec l'Asie avant la Compagnie des Indes Orientales de Colbert," Histoire Généalogie, 2003 at http://www.histoire-genealogie.com/spip.php?article158.

Le Lan, J-Y., « Le Naufrage du Compte de Maurepas, » Histoire-Généalogie, at http://www.histoire-genealogie.com/article.php3?id_article=168, November 1, 2004.

Maurel, H., « Navires de la Compagnie des Indes allant aux/venant des Isles de l'Océan Indien, » at http://perso.orange.fr/henri.maurel/cieind2.htm.

Milmo, C., "Pirates of the Channel Islands: A £200m treasure hunt," The Independent, June 12, 2008 at http://www.independent.co.uk/news/uk/home-news/pirates-of-the-channel-islands-a-163200m-treasure-hunt-845054.html.

Rockel, E., "De l'île sans nom à l'île Bourbon ou, Bourbon des origines jusqu'à 1700," at http://amis.univ-reunion.fr/Conference/Complement/166_bourbon/index_bourbon.html.

Rockel, E., « Vieux gréements—Le Soleil d'Orient, » at http://www.locmiquelic.org/Bateaux/index.html.

Sulliman, "Bourbon au secours des Indes," Clicanoo, le journal de l'ile de la Réunion, March 13, 2005, at http://www.clicanoo.com/article.php3?id_article=99074.

Vacher, P., « Contribution à l'histoire de l'établissement français à Madagascar par le Baron de Benyowszky (1772-1776), D'après de nouvelles sources manuscrites », Collection 'Clio en Afrique,' N° 19, été 2006, Editions du Centre d'Etude des Mondes Africains, MMSH, Aix en Provence, at http://www.mmsh.univ-aix.fr/iea/Clio/VACHER.pdf.

Vaxelaire, D., Service Multimédia, March 4, 2005, at http://reunion.rfo.fr/article124.html.

Word of Wordland, « Count Matus Moric Benovsky », at http://www.slovakopedia.com/m/moric-benovsky.htm.

Pirates

Bastions Pirates, Do or Die, Ed. Aden, No. 8 (2001), reproduced at http://www.eco-action.org/dod/n08/pirate.html and a French translation at http://mathieu.saura.free.fr/site/texts/text%20bastions%20pirates.htm.

Captain William Kidd, USS KIDD Veterans Memorial, Louisiana Naval War Memorial Commission, 2006, at http://www.usskidd.com/willkidd.html.

Durup, J., "Short seafaring adventures and conflicts in the Indian Ocean 1405-1811," 2004, at http://perso.orange.fr/henri.maurel/seafaring%201.htm.

Hawkins, P., "The Ultimate Captain William Kidd website" at http://www.Captain kidd.pwp.blueyonder.co.uk/.

Kingston, W.H.G. and Frith, H., Notable Voyagers, Chapter XXII, Dampier's voyages, at http://www.athelstane.co.uk/kingston/voyagers/vyage22.htm.

Ossian, R., "Pirate's Cove," at http://www.thepirateking.com/bios/chivers_dirk.htm.

Pirate Encyclopedia, at http://ageofpirates.com/.

Privateers Dragons of the Caribbean, at http://www.privateer-dragons.com/

Rule, C., "Piratical History of Madagascar," in Pirate Strongholds & Hideouts, http://www.piratesinfo.com/detail/detail.php?article_id=70.

s.n., A Buccaneer's Atlas, Basil Ringrose, at http://content.cdlib.org/xtf/view?docId=ft7z09p18j&chunk.id=d0e1879&toc.depth=1&toc.id=d0e1720&brand=eschol.

s.n., Captain Swan at http://www.burleygames.com/Captain swan.htm.

s.n., William Dampier, NNDB, at http://www.nndb.com/people/943/000096655/.

s.n., « Olivier Le Vasseur, dit La Buse, le Pirate, » at http://dossiers.clicanoo.com/article.php?id_article=97577&id_mot=.

Stapleton, D., "Pirate Roster—Olivier La Bouche », 2001, at http://pirateshold.buccaneersoft.com/roster/olivier_la_bouche.html;

Vallar, C., "Black Pirates," in Pirates and Privateers, The History of Maritime Piracy, at http://www.cindyvallar.com/blackpirates.html.

Wikipedia, Charles Swan, at http://en.wikipedia.org/wiki/Charles_Swan.

British Royal Navy

Michael Phillips' ships of the old Navy, at http://www.ageofnelson.org/MichaelPhillips/info.php?.

Sailing ships of the Royal Navy, http://www.cronab.demon.co.uk/;

s.n., "Sir Thomas Troubridge," at http://www.aboutnelson.co.uk/13troubridge.htm.

French Navy

Gerbeau, H., "Des minorités mal connues : esclaves indiens et malais des Mascareignes au XIXe siècle," in Table ronde sur "Migrations, minorités et échanges en Océan Indien, XIXe-XXe siècle," Sénanque, 1978, *Études et Documents*, Aix-en-Provence, IHPOM (Institut d'Histoire des Pays d'Outre-Mer), Université de Provence, n° 11, 1979, p. 160-242, at http://classiques.uqac.ca/contemporains/gerbeau_hubert/minorites_mal_connues/minorites_mal_connues.html.

Musée Lapérouse, Albi, « Découvrir, Navires en service dont le nom évoque Lapérouse », at http://www.mairie-albi.fr/arthisto/gens/navires.html

s.n., "I have not yet begun to fight!" at http://www.hoala.org/Grade_7/jones.html.

The Serapis project, at http://www.serapisproject.org/.

Wikipedia, "HMS Serapis" at http://en.wikipedia.org/wiki/HMS_Serapis_(1779).

Sailing Vessels and Steamers of the 19th Century

Clydebuiltships, Shipping Times, SS Peshawur, at http://www.clydesite.co.uk/clydebuilt/viewship.asp?id=15220

The Red Duster Website at http://www.red-duster.co.uk/UNION11.htm

White Star Line ships at http://www.titanic-whitestarships.com/.

Russian Fleet 1904-1905

Cobb, J., "Russo-Japanese War, 1904-1905: A Naval Perspective (Part 2)," at http://www.gamesquad.com/distantguns/2006/02/21/russo-japanese-war-1904-%E2%80%93-1905-a-naval-perspective-%E2%80%93-part-2/.

The Russo-Japanese War Research Society, at http://www.russo-japanesewar.com/naval_links.html.

Compagnie des Messageries Maritimes

Web site of the Messageries Maritimes at http://www.messageries-maritimes.org/.

Compagnie Havraise Péninsulaire

Miramar ship index, at http://www.miramarshipindex.org.nz/ship/list.

s.n., « Histoire de Tamatave » at http://tamatave.ifrance.com/histoire/histoire.htm.

s.n., « Perte du Ville de Majunga » at http://naviresdeguerre.free.fr/majunga.html

WWII—Diego Suarez

Butler, M., "The British Loyalty," 2002, at http://www.gan.philliptsmall.me.uk/Articles/BritishLoyalty.htm.

Daubigny, B., "Le Drame de Diégo-Suarez," at http://perso.orange.fr/bertand.daubigny/Hdiego.htm.

Godin, "Historique du Premier *Beveziers*" Actes du XXVè congrès des Sociétés Historiques et Archéologiques de Normandie (October 4-7, 1990), communications de Vergé-Franeeschi, M. and Zesberg, A., at http://sous-marins.chezalice.fr/historique_du_sous_marins_a_doub.htm.

Hackett, B. and Kingsepp, S., "Sensuikan!," at http://www.combinedfleet.com/Madagascar.htm.

s.n., "Les batiments ayant porté le nom de Victor Schoelcher," at http://www.netmarine.net/bat/ae/schoelcher/ancien.htm.

Turner, L., "Derrick and Sunderland T9040," Royal Air Force Archive List No. A5900861, at http://www.bbc.co.uk/ww2peopleswar/stories/61/a5900861.shtml.

Other Wreckages

The Red Duster Website at http://www.red-duster.co.uk/UNION11.htm.

http://fortships.tripod.com/war_damagelosses.htm

http://uboat.net/boats/u197.htm.

Index

Abreu, Aleixo de 30, 33,
Aceh (Indonesia) 62, 113, 114
Aden (Yemen) 173, 174, 283, 286, 299
Aenes Francês, Pêro (Captain) 34
Albuquerque (de), Alfonso (Captain) 25, 27, 29
Allaire, Charles and Jean 121
Amaral, Manuel Bothelho do (Captain) 42
Amber Cape 2, 28, 29, 51, 189, 274, 284
Ambohimanga 7
Amsterdam, Holland 58, 60, 63, 64, 67, 70, 72, 75, 78, 93, 118, 144
Andriamamory (Chief) 66, 67
Andriamasikara, King of Anosy 72
Andrianampoinimerina, King of Madagascar 7
Andrian-Kirinda, King of Androy 91, 94
Andrian-Tsiambany, King of Anosy 67
Androy 72, 91, 93, 94
Ango, Jean 113, 114
Angoza island (Mozambique) 28, 29
Anjouan (Comoros) 97, 106, 134, 155, 156, 163, 183, 184, 185
Annet, French Governor General Armand 285
Anosy 19, 52, 67, 72, 90, 93, 111, 119, 165
Anquez 210, 211, 216
Antananarivo 7, 95, 213, 227, 230, 234, 263, 284, 285, 290
Antongil Bay 7, 10, 60, 62, 63, 68, 73, 74, 78, 82, 119, 120, 122, 124, 127, 138, 140, 141, 142, 143, 154, 190, 223, 224, 225, 295, 296
Antwerp, Belgium 56, 265
Après de Mannevillette, Jean-Baptiste d' 288
Augustine (St.), Bay 3, 30, 33, 34, 38, 41, 42, 52, 60, 82, 83, 84, 91, 94, 95, 99, 101, 102, 105, 112, 114, 130, 144, 145, 154, 158, 163, 164, 176, 177, 179, 180, 187, 194
Averill, Cyrus B. (Captain) 228, 229, 230
Avery, Henry (Pirate Captain) 157, 173

Baba, King of Tulear 105
Baldridge, Adam 164

Bantam (Indonesia) 60, 62, 64, 66, 133, 200
Barbados 3
Barbotin, François (Captain) 139
Barlow, Edward (Captain) 168
Baron (Captain) 130, 131
Barracouta Islands 262, 292
Barren Islands 114
Bartels, KrvKpt. Robert 299
Batavia (Jakarta), Indonesia 42, 69, 70, 74, 75, 76, 146, 181, 200
Bazoche (Commander) 280
Beauregard, de (Captain) 127, 128, 129
Beausse, Pierre de 121, 122
Beauvais, Jean Le Fer de (Captain) 289
Belgium 56
Bellamy, Samuel (Pirate Captain) 184
Bellard (Captain) 229
Bellomont, Richard Coote, earl of (Governor of New York) 171
Benbow, John 93
Bengal, India 82, 89, 90, 96, 99, 144, 148, 163, 202, 288
Benyowsky, Count Maurice A. 141, 142, 143
Bissell, Austen (Captain) 199, 200
Blanquet de la Haye, Jacob (Admiral) 127
Boeny (Bay) 2, 25
Boispéan, de (Captain) 133
Boispréaux, Garreau de (geographer) 142
Bollan, Jacques de 74
Bombay, India 88, 89, 96, 97, 134, 158, 174, 175, 183, 228, 236
Booth, George (Pirate Captain) 177, 178, 179, 194
Boston 164, 171, 180, 198, 228, 234, 239
Both, Pieter 66
Boulanger (Captain) 135

Bourbon (Reunion) 6, 117, 119, 121, 125, 128, 129, 131, 134, 135, 136, 137, 138, 139, 140, 148, 177, 181, 183, 190, 195, 197, 198, 202, 208, 222, 287, 288
Bourdonnais, Bertrand Mahé, comte de la, 137, 138
Bowen, John (Pirate Captain) 90, 91, 111, 177, 178, 179, 180, 194
Brazil 2, 40, 42, 113, 124, 125, 126, 178, 183
Bretèche, de la 111, 129, 130
Bucquoy, Jacob de 187, 191
Burgess, Samuel (Pirate Captain) 94, 176, 177, 181, 182, 193

Caen, France 115
Calcutta, India 89, 168, 174, 236
Camara, Jose Pedro da 44
Campbell, John. P. (American consul) 235
Cape of Good Hope 14, 16, 22, 23, 37, 57, 58, 60, 69, 70, 71, 73, 74, 75, 81, 82, 90, 97, 99, 113, 121, 125, 134, 145, 146, 153, 157, 167, 171, 183, 198, 199, 200, 201, 206, 271, 289, 299
Cape Town 72, 142
Cape Verde 70, 121, 183, 219
Caraccioli (Monk) 156, 157
Caron, François 117, 125, 126, 146
Castillon, Antoine-Paul de (Captain) 287
Castro, Fernando de (Friar) 41
Cauche, François 72, 146
Caunay, Pierre (Captain) 113
Charles II, King of England 83
Charles V, King of Spain 55
Châtelain, Françoise 131
Cheng Ho 2
Chivers, Dirk (Pirate Captain) 164, 171, 173, 174, 176, 177
Christian, Rear Admiral Sir Hugh 198

Churcher, John (Captain) 180
Churchill, Winston 273
Claëssens, Nicolas (Captain) 139
Clifford, Barry 171
Clive, Robert 117
Cochin (India) 22, 41, 168
Cockburn (Captain) 158
Coin, Jan (Captain) 76, 192
Colbert, Jean-Baptiste 117, 120, 124, 126, 131
Colon (Captain) 120
Comoros Islands 28, 40, 82, 88, 97, 114, 134, 151, 156, 163, 167, 173, 176, 181, 184, 222, 227, 254, 268, 280
Condent, Christopher (Pirate Captain) 171, 182, 183, 195
Cook, James (explorer) 141
Cook, John (Captain) 161
Cooper, Lames (Sir) 75
Coquet (Captain) 119
Cornuël (Captain) 122
Corrientes, Cape 81
Costa, Alfonso Lopes da (Captain) 29
Costa, Garcia da (Captain) 29
Cottineau de Kerloguen (Captain) 206
Coulon, Nicole 131
Courrier Bay 27, 274, 277, 281
Courteen, William 3
Covilh, Pro da 22
Craig, Lieutenant Henry 202
Cromwell, Oliver 83
Culliford, Robert (Pirate Captain) 165, 171, 173, 174, 175, 176, 177, 193
Cunha, Nuno da 30, 33, 34, 36, 38, 40, 52, 53
Cunha, Pêro Vaz da (Captain) 36, 38
Cunha, Simão da (Captain) 36
Cunha, Tristan da 23, 25, 27, 28, 29, 50, 53

Dale, John 99, 101, 102, 105, 106
Dampier, William 160, 161, 162, 163, 192
Daniel (Captain) 233
Davis, Edward (Pirate Captain) 160, 161
Davis, Howell (Pirate Captain) 184
Defoe, Daniel 18
De Gaulle, General Charles 273, 285
Delauze, Henri-Germain 134
Delaware 69
Delort (Commander) 281
Depanis, Lieutenant de vaisseau Hyppolite 208, 209
des Essarts, Jacques (Captain) 140
Desforges Boucher, Governor of Bourbon 183, 190
Desjardin (Captain) 288
Desmousseaux (Captain) 142
Dias, Bartolomeu 22
Dias, Diogo 22
Diego Suarez (Antsiranana) 7, 156, 242, 256, 257, 259, 266, 268, 271, 273, 274, 277, 279, 280, 281, 282, 283, 284, 285, 292, 293, 298
Dieppe, France 33, 72, 113, 114, 115, 117, 118, 119
Digart, Grégoire (Captain) 118
Dijk, Jan Willemszoon (Captain) 68
Dobrotvorskii, L. F. (Captain) 242, 243
Douglas, John Erskine (Captain) 198
Douweszoon, Gerard (Captain) 68
Drake, Sir Francis 152
Drevet (Captain) 258
Drummond, Robert (Captain) 91, 93, 112
Drury, Robert 18, 91, 93, 94, 95
Dumont d'Urville, Jules 207
Dundas, George (Captain) 97, 99, 101, 102
Dupleix, Joseph François 117

Durand-Linois, Admiral Charles de 199

Edgecumbe (Captain) 174, 175
Elizabeth I, Queen of England 56, 81
England, Edward (Pirate Captain) 184, 185
Enkhuizen, Holland 63, 66, 70
Ericeira, Don Luis de Meneses 185

Fabré, Théodore (Captain) 208
Falconer, William 96
Fanning, Nathaniel 206
Farnese, Alexander, Duke of Parma 56
Farrell, Joseph (Captain) 173
Feljkerzam, Counter-Admiral D.G. 242, 243
Félix, Paul (Captain) 292
Fenerive (Fenoarivo Atsinanana) 11, 73, 143, 177, 179, 181, 299
Flacourt, Etienne de 115
Fletcher, Benjamin (Governor of New York) 112, 164
Fonseca, Diogo da (Captain) 33, 36, 114
Fonseca, Duarte da (Captain) 33
Fonseca, Lucas da (Captain) 29
Fontaine (Capitaine de frégate) 277
Forde, Colonel Francis 96
Forgeard (Captain) 264, 265
Fort Dauphin 2, 3, 11, 16, 17, 18, 29, 72, 74, 75, 76, 90, 93, 95, 115, 120, 121, 122, 124, 125, 126, 127, 128, 129, 130, 131, 134, 135, 139, 141, 142, 143, 144, 164, 198, 208, 214, 221, 222, 259, 268, 291, 292, 295
Foulpointe (Mahavelona) 11, 120, 121, 122, 126, 127, 137, 138, 139, 142, 143, 181, 208, 262, 287, 288, 289, 297, 298

Foucquembourg 119
Fouquet, Viscount Nicolas 117, 146
Franco-Hova Wars 7
Franklin, Benjamin 143, 206
Franszoon, Jacob (Captain) 66
Freke, William (Captain) 84, 87, 88, 89

Gallieni, General Joseph 213, 261
Gama, Vasco da 22
Gijsbertsz., Theunis (Captain) 74
Gilbert (Captain) 259
Girardin (Captain) 121
Glover, Robert (Captain) 173
Goa (India) 22, 29, 30, 34, 40, 42, 43, 44, 66, 89, 185
Gomes de Abreu, João (Captain) 28, 29, 50
Gooyer, Cornelis Simonsz. 118
Goubert, Alonse (Captain) 117, 118, 119
Goubeyre, Capitaine de vaisseau Jean-Marie 208
Gout, Jean-Pierre (Captain) 213
Grammont, de (Captain) 176, 177
Grosos, Eugène 261

Hallstrom, Sverker 134
Halsey, John (Pirate Captain) 91, 178, 180, 181, 182, 183
Harland (Captain) 155
Harris, Peter (Pirate Captain) 160
Hay, Michel (Captain) 140
Heinszoon, Kornelis (Vice Admiral) 60
Henri III, King of France 113
Henri IV, King of France 115
Henry the Navigator (Prince) 21
Hentsch, Edouard 261
Hirst, Reverend William 96
Hitler, Adolf 272, 273
Hoar, John (Pirate Captain) 164, 173, 174

Holland 56, 60, 67, 69, 70, 73, 75, 76, 77, 115, 268
Hoorn, Holland 69, 70
Hornigold, Benjamin (Pirate Captain) 184
Houtman, Cornelis de 58
Hova 212, 224, 225, 227, 234, 235, 290, 291
Howard, Thomas (Pirate Captain) 112, 178, 179, 180
Huguet (Captain) 213, 214, 215

IJsbrandszoon, Jakob (Captain) 66
Inless, Samuel (Pirate Captain) 174, 177

Jacobs, Thomas J. 220
Jacomy-Régnier 16
Jahangir, Mughal emperor 82
James (Pirate Captain) 178, 179
James I, King of England 81
James II, King of England 160
Janszoon, Willem (Captain) 64
Jesus, Athanase de (Friar) 66
João II, King of Portugal 22
João III, King of Portugal 33, 34, 36
Joffre, General Joseph 274
Johnson, Charles (Captain) 18, 180, 182, 183
Jones, Achen (Pirate Captain) 176,
Jones, John Paul (Captain) 205, 206
Jong, Maximilien de 73
José, King of Portugal 44

Karimboly, coast 66, 67, 70
Kelley, James (Captain) 175
Kendall, Abraham (Captain) 81
Kergadiou (Captain) 121, 122
Kidd, William (Pirate Captain) 154, 165, 166, 167, 168, 170, 171, 173, 175, 176, 193
Klunt, Jakob Jakobszoon (Captain) 63
Knight (Pirate Captain) 88, 163

Kynnaston, Thomas 3

Labeda (Captain) 132
Laborde, Jean (French Consul) 227, 290
Labriants (Captain) 120
La Buse (Le Vasseur, Olivier) (Pirate Captain) 158, 184, 185, 186, 187, 188, 189, 190, 195
La Butte-Frérot, Guillaume de (Captain) 135, 136
Laccadive Islands (India) 26, 168, 175, 185
Lacerda, Manoel de (Captain) 30, 33
La Clocheterie, de (Captain) 121
la Forest des Royers (Captain) 117
La Hure, de 128
Lambert, Joseph-François 7
Lancaster, James (Captain) 81, 82
Lange, Dirk de (Captain) 75
Laréquier, Albéric (Captain) 254
La Roche-Saint-André, de (Captain) 120
La Rochelle, France 119, 124, 132
Laut, Raja (Ruler of Mindanao, Philippines) 162
La Vigne, de (Captain) 122
Law, John 117
Le Bourg (Captain) 115
Lecoole, Jerrey (Captain) 184
Lee, Thomas (Captain) 96
Legall (Captain) 299
Lemaire (Commander) 280, 281
Le Meme, François (Captain) 289
Lemos, Duarte de (Captain) 29
Lenepveu (Captain) 217
Le Vasseur, Olivier : see La Buse
L'Herrec (Captain) 268
Lheritier 206
L'Hermitte (Captain) 190
Libertalia 7, 156, 157
Lima, Antonio de (Captain) 29
Limà, Fernando de (Captain) 36, 38
Lintgens, Pieter 115

Littleton (Commodore) 154, 155
Livingston, Robert (Colonel) 165, 166
Lorient, France 132, 135, 138, 139, 140, 141, 142, 144, 287, 288, 289
Louis XIII, King of France 17, 115
Louis XIV, King of France 115, 117, 131, 133, 134
Louis XV, King of France 139, 141, 215
Lucas, Fred 230
Lyndsay, Sir John 97

Mackraw (Captain) 185
Madras, India 99, 106, 138, 175, 181, 199, 200
Maerten (French naval commander) 279, 280, 281
Magon de Medine, Charles René (Captain) 289
Mahajamba (Bay) 2, 27
Mahajanga (Majunga) 2, 25, 30, 179, 189, 190, 213, 227, 236, 254, 280, 285
Malabar (coast), India 88, 111, 158, 168, 174, 177, 180, 194
Maldives 2, 82, 114, 184
Malindi (Kenya) 2, 23, 28, 40
Mallet, Jean (Captain) 106, 289
Manambolo (river) 114
Manambovo (river) 3, 71
Mannberguer, Frédéric 261
Manoel, King of Portugal 28
Maroantsetra 141
Martin, François 121, 126
Mascarenhas, Pêro 34
Masson (Captain) 266
Matitanana 11, 23, 27, 28, 29, 33, 95, 119, 176, 181, 292
Matthews, Thomas (Commodore) 158, 187, 189
Maudave, Louis Amédée Fayd'herbe, Comte de 141

Maurice of Nassau, Prince of Holland 56, 76
Mauritius 3, 6, 58, 60, 62, 66, 73, 74, 75, 76, 82, 83, 90, 106, 117, 118, 119, 135, 137, 138, 139, 140, 141, 142, 143, 144, 145, 178, 179, 185, 197, 198, 201, 202, 207, 217, 219, 222, 235, 236, 256, 263, 268, 287, 288, 289, 290, 291, 300
May, William (Captain) 175
Maybury, Lt. Commander S.L.B. 277
Mayotte (Comoros) 173, 179, 184, 209, 222
Meilleraye, Marshal Duke Charles de la 117, 119, 120
Mendoza, Christovão de (Captain) 30
Merina (kingdom) 6, 7, 8, 217
Middleton, David (Captain) 82
Middleton, John (Captain) 82
Miller, James (Captain) 181
Miot, Admiral Paul-Emile 227
Misson, James (Captain) 7, 156, 157
Moheli (Comoros) 156, 163, 167
Moluccas, Indonesia 113, 114
Mombassa (Kenya) 2, 23, 40, 283
Mondevergue, François de Lopis, marquis de 125, 126
Montaubon 122
Moors 55, 90, 91, 154, 177
Morel (Captain) 141
Morgan, Sir Henry 158
Morondava 84, 87, 88, 135, 163, 214, 230, 234
Morphey, Nicolas (Captain) 139
Morrell, Benjamin (Captain) 218, 219, 220, 221, 222
Morrison, James 200, 201
Mozambique 11, 22, 23, 25, 27, 28, 29, 33, 37, 40, 42, 44, 66, 87, 105, 106, 114, 126, 130, 131, 133, 142, 187, 188, 189, 222

Mozambique Channel 1, 16, 22, 23, 25, 29, 30, 58, 81, 87, 97, 99, 127, 130, 156, 189, 259, 261, 271, 300
Muterse de Guérande, Jean (Captain) 144, 145

Napoléon III, Charles Louis, Emperor of France 7
Naraï, King of Siam 133, 134
Nebogatov, Counter-Admiral Nicholas I. 243
Nevé (Captain) 140
New York 164, 165, 167, 171, 175, 177, 218, 219, 220, 221, 228, 291
Nicholas II, Tsar of Russia 241
Nicobar Islands (India) 163, 175, 180
Noboru, Rear Admiral Ishizaki 283
North, Nathaniel (Pirate Captain) 174, 176, 177, 181, 194
Nosy Be (island) 3, 83, 214, 241, 242, 243, 244, 254, 280, 297
Nosy Boraha: see Island of St. Marie
Nosy Hao (Murder Island) 178, 194
Nosy Mangabe (Marosy Island) 138, 141
Nosy Manitsa (island) 60, 70
Nosy Manja (islet) 2, 27

Oman 88, 184
Orcel (Captain) 209
Orléans, Philippe Duc d', Regent of France 137

Paiva, Gaspar de (Captain) 30
Parmentier, Jean (Captain) 114
Parmentier, Raoul (Captain) 114
Pasley, Admiral Sir William 198
Pearson, Mike Parker 91
Pearson, Richard (Captain) 205
Pellegrini (Captain) 298
Pellew, Sir Edward 199, 200, 201

Perreira de Coutinho, Rui (Captain) 23, 50
Persia 122, 126, 151
Persian Gulf 1, 2, 7, 154, 164, 165
Philip II, King of Spain 55, 56
Phillips, Frederick 177, 193
Pierre, Rear Admiral 227
Pigot, George (Captain) 200
Pitcairn, Robert 96
Plowman, Daniel (Captain) 180
Poivre, Pierre (Governor of Mauritius) 289
Pondicherry, India 137, 138, 140
Preneyre (Captain) 259
Pro, John (Pirate Captain) 155
Pronis, Jacques 115, 119
Proté (Captain) 222, 223

Quiloa (Tanzania) 29, 30

Radama I, King of Madagascar 7, 208
Radama II, King of Madagascar 7, 227
Raeder, Grand Admiral Erich 272
Rainier, Peter (Captain) 200
Rainilaiarivony, Prime Minister 227, 230
Ranavalona I, Queen of Madagascar 7, 208, 213, 224, 290
Ranavalona II, Queen of Madagascar 227
Ranavalona III, Queen of Madagascar 227, 230
Ranofotsy Cove (Galion's Bay) 3, 33, 34, 52, 114
Ramboasalama 7
Rasoherina, Queen of Madgascar 227
Ravisseau (Captain) 143
Raymond, George (Captain) 81
Read, John (Pirate Captain) 162, 163
Rebours (Captain) 262
Red Sea 1, 7, 72, 74, 88, 111, 117, 118, 119, 120, 122, 146, 154,

157, 158, 164, 165, 168, 173, 174, 176, 177, 179, 180, 181, 183, 194
Régimont, Gilles de (Captain) 72, 115
Regnault, Etienne 121, 128
Reunion 6, 37, 115, 117, 118, 121, 183, 185, 187, 190, 201, 202, 212, 213, 225, 227, 253, 256, 262, 263, 266, 268, 281, 290, 293
Reyetsz, Cornelis (Captain) 68
Richard (Lieutenant de vaisseau) 277
Richards (Commodore) 155
Richelieu, Cardinal Armand Jean de 115, 117, 119
Richmond (Captain) 120
Ridge, John James (Captain) 201
Rigault (Captain) 115
Ringrose, Basil 158, 160, 161
Robinson, W. W. (American consul) 234
Roche, Lieutenant de Vaisseau 206
Rodrigues Island 119, 202
Rogers, Woodes (Governor of the Bahamas) 158, 182, 184, 195
Rommel, Fieldmarshal Erwin 273
Roosevelt, Theodore, American President 273
Rostand, Albert 249
Rotterdam, Holland 74
Rouen, France 115
Roux, Sylvain 208
Rowley, Josias (Commodore) 197
Roy, Girard de 115
Rozhdestvenskii, Vice-Admiral Zinovi P. 241, 242, 243, 244

Sadia 3
Saint Claire (Itaperina) 118
St. Helena, Island 69, 72, 82
St. Luce, Bay (Manafialy) 33, 34, 60, 64, 67, 68, 72, 77, 115, 118, 119, 198

St. Malo, France 114, 115, 183, 225, 288
St. Marie, Cape 16, 144
St. Marie, Island 7, 11, 60, 73, 74, 82, 95, 120, 121, 122, 126, 138, 139, 140, 154, 157, 158, 164, 170, 174, 177, 179, 183, 189, 190, 193, 206, 213, 223, 224, 225, 227, 288, 290, 297
Sakalava (Kingdom and tribe) 94, 142, 230, 234
Salara 44, 101, 228, 230
Saldanha, Antonin da (Captain) 28
Saldanha, António de (Captain) 30, 36, 38
Sampayo, Lopo Vaz de 34
Samuel, Abraham (Captain and King of Anosy) 90, 93, 111, 164
Sanglier, Chevalier de 143
Sanguinet, Jean-Joseph de (Captain) 139
Saunier, de (Captain) 142, 143
Sawbridge (Captain) 173
Scheffino (Captain) 291
Scrafton, Luke (Esq.) 96
Secord, Arthur W. 91
Séchelles, Jean Moreau de 139
Sequeira, Diogo Lopez de (Captain) 29, 52
Seychelles 88, 139, 184, 197, 281, 289
Silva, Balthazar da (Captain) 30
Silva da Gama, Pêro da (Captain) 40
Silveira, João da (Captain) 29
Simons, Ernest 249
Smuts, Field Marshal Jan C. 273
Snow, Raymond (Captain) 97
Soares, Diogo (Captain) 40, 41
Soares, Ruy (Captain) 29
Socotra (off Yemen) 33, 122, 126
Sofala (Mozambique) 11, 105, 163
Soulas, Jacques 118, 119
Sousa de Campelo, Hermogénio de (Captain) 43

Sousa, Diogo Lopes de (Captain) 40
Sousa "Mancias", Francisco de (Captain) 29, 30
Sousa, Martim Alfonso da 40
Sousa, Pêro Lopes da (Captain) 40
Spens, Nathaniel 101
Stanwood, Victor F. W. (American consul) 230, 234, 235
Star Bank (Banc de l'Etoile) 76, 88, 97, 127, 144
Sténuit, Robert 44, 45, 54, 106, 134
Stewart, Alexander (Captain) 91, 112
Stokmans, Pieter (Captain) 62
Stout, Ralph (Pirate Captain) 174, 175
Street, Lieutenant Benjamin 201
Sumatra, Indonesia 58, 62, 113, 114, 162, 163
Sunda Straight 58, 221
Surat (India) 42, 82, 83, 111, 126, 127, 130, 131, 133, 134, 140, 164, 168, 173
Swan, Charles (Pirate Captain) 158, 160, 161, 162
Swete, Dick 207
Syfret, Rear-Admiral Sir Edward Neville 273

Taaikaas, Jan (Captain) 70, 71
Tamatave 11, 208, 213, 215, 217, 222, 225, 227, 233, 234, 236, 256, 262, 263, 264, 265, 285, 290, 291, 292, 293, 294, 295, 296, 297, 298
Taylor, John (Pirate Captain) 158, 183, 184, 185, 186, 187, 188, 189, 195
Teat, Josiah (Pirate Captain) 88, 161, 162, 163
Telos, Alvaro (Captain) 23
Tew, Thomas (Pirate Captain) 156, 157
Texel, Holland 58, 60, 62, 63, 66, 68, 69, 72, 206
Thomaszoon, Jan (Captain) 69
Townley (Pirate Captain) 160
Tripier de Barmont, Antoine (Captain) 288
Tromelin, chevalier de 143
Trott, Thomas Witney (Captain) 299
Troubridge, Edward Thomas 200, 201
Troubridge, Vice Admiral Sir Thomas 199, 200
Truchot de La Chesnaye (Captain) 121, 122
Tsimanangarivo, King of Sakalava 94
Tsingilofilo Bay 41, 51
Tulear (Toliary) 3, 41, 42, 60, 105, 214, 229, 230, 285
Tuttle, Michael 207

van der Haghen, Steven (Admiral) 60
van der Stel, Adriaan 73, 118, 119
van Neck, Jacob (Admiral) 60
van Oldebarnevelt, Johan 56
van Riebeeck, Jan 72, 73, 78
van Spaan, Gerrit 14
Vansittart, Henry 96
Vasconcelos, Luis Fernandes de (Captain) 41
Vaz o Roxo, Pêro 34, 53
Vega, João da (Captain) 28, 50
Verburgh, Frederik 72, 73, 74
Vergé, L. du 235
Véron (Captain) 121, 122
Verrazano (brothers) 113
Villebague, Julien de la (Captain) 288
Virginia 69, 95, 161, 178
Visser, Koert Geurtszoon (Captain) 69
Vlamingh, Willem de 75

Vohemar 42, 124, 127, 251, 257, 294, 295
Volkertsz, Samuel (Captain) 72
Vries, David Pieterszen de 69

Warren (Commodore) 154, 174
White, Thomas (Pirate Captain) 179, 181, 194
Wielingen, Holland 70, 75
Wilhelm, Casimir (Captain) 225, 227
William III, King of England 165
William of Orange 56

Williams, David (Pirate Captain) 155, 176, 177, 179, 181, 182, 193
Wood, James Athol (Captain) 198
Woodworth, Selim E. 220, 222
Woolley (Captain) 179
Wright (Captain) 168

Young, Nicholas (Captain) 90
Young, William (Captain) 89, 90, 180

Zanzibar 23, 40, 145, 178, 179, 228, 254, 283

About the Author

Pierre van den Boogaerde was born and raised in Ghent, a provincial town in northwest Belgium. He studied law and economics in Belgium and then earned an MBA degree at the University of Chicago. After working for a commercial bank in New York and then in Brussels, he spent most of his career working for an international financial organization based in Washington, D.C. He took up several overseas assignments, including Egypt, Côte d'Ivoire and Madagascar, where he was posted about three years ago.

Pierre discovered diving while residing in Egypt, about a quarter of a century ago. One of his early dives was on the *Thistlegorm,* a WWII wreck in the Red Sea, which still contains trucks, motorcycles and airplane parts in the hull. That dive got him hooked. Since then, Pierre has dived wrecks on several continents. While residing in Washington D.C., he was a frequent visitor to the *Graveyard of the Atlantic* in the area of Beaufort, N.C.

Amongst his early outings in Madagascar, Pierre dove the *Winterton* and *Nossa Senhora do Monte do Carmo* wrecks off Salara. He immediately wanted to know more about wrecks off Madagascar, but soon realized that nothing had ever been published about it. This led him to start researching the subject. Thousands of hours of research and writing, and many dives later, it all ended up in this book.

Printed in the United Kingdom by
Lightning Source UK Ltd., Milton Keynes
138632UK00001B/104/P